CONTENTS

ACKNOWLEDGMENTS

This book has benefitted from discussion and collaboration with many people. I am particularly indebted to the members of my lab, who have struggled alongside me through some of these ideas, and to several people who gave comments on parts of the book: Rahul Bhui, Fiery Cushman, Anatole Gershman, and Rebecca Saxe. Finally, I am grateful to my mother Nancy, for teaching me how to write, my father Anatole, for teaching me how to think, and my wife Anna, for teaching me how to argue.

What Makes Us Smart

1

Introduction

ARE WE SMART?

Against stupidity the gods themselves contend in vain.
—FRIEDRICH SCHILLER

This book is motivated by a fundamental puzzle about human cognition: How can we apparently be so stupid and so smart at the same time? On the one hand, the catalog of human error is vast: we perceive things that aren't there and fail to perceive things right in front of us, we forget things that happened and remember things that didn't, we say things we don't mean and mean things we don't say, we're inconsistent, biased, myopic, overly optimistic, and—despite this litany of imperfections—overconfident. In short, we appear to be as far as one can imagine from an ideal of rationality.[1]

On the other hand, there is an equally vast catalog of findings in support of human rationality: we come close to optimal performance in domains ranging from motor control and sensory perception to prediction, communication, decision making, and logical reasoning.[2] Even more puzzlingly, sometimes the very same phenomena appear to provide evidence both for and against rationality, depending on the theoretical lens through which the phenomena are studied.

This puzzle has been around for as long as people have contemplated the nature of human intelligence. It was aptly summarized by Richard Nisbett and Lee Ross in the opening passage of their classic book on social psychology:

One of philosophy's oldest paradoxes is the apparent contradiction between the great triumphs and the dramatic failures of the human mind. The same organism that routinely solves inferential problems too subtle and complex for the mightiest computers often makes errors in the simplest of judgments about everyday events. The errors, moreover, often seem traceable to violations of the same inferential rules that underlie people's most impressive successes.[3]

As indicated by Nisbett and Ross, the puzzle of human intelligence is reflected in our conflicted relationship with computers. On the one hand, it has long been advocated that error-prone human judgment be replaced by statistical algorithms. In 1954, Paul Meehl published a bombshell book entitled *Clinical Versus Statistical Prediction*, in which he argued (to the disbelief of his clinical colleagues) that the intuitive judgments of clinical psychologists were typically less accurate than the outputs of statistical algorithms. This conclusion was reinforced by subsequent studies and expanded to other domains.[4] For example, in his 2003 book *Moneyball*, Michael Lewis popularized the story of the baseball manager Billy Beane, who showed (to the disbelief of his managerial colleagues) that statistical analysis could be used to predict player performance better than the subjective judgments of managers.[5] Today, the idea that computers can outperform humans, even on tasks previously thought to require human expertise, has become mundane, with stunning victories in Go, poker, chess, and Jeopardy.[6]

And yet, despite these successes, computers still struggle to emulate the scope and flexibility of human cognition.[7] After the Go master Lee Sedol was defeated by the AlphaGo computer program, he could get up, talk to reporters, go home, read a book, make dinner, and carry out the countless other daily activities that we do not even register as intelligence. AlphaGo, on the other hand, simply turned off, its job complete. Even in the domains for which machine learning algorithms have been specifically optimized, trivial variations in appearance (e.g., altering the colors and shapes of objects) or slight modifications in the rules will have catastrophic effects on performance. What seems to be missing is some form of "common sense"—the set of background beliefs and inferential abilities that allow humans to adapt, almost effortlessly, to an endless variety of problems.

The lack of common sense in modern artificial intelligence (AI) systems is vivid in the domain of natural language processing. Consider the sentence "I saw the Grand Canyon flying to New York."[8] When asked to translate into German, Google Translate returns "Ich sah den Grand Canyon nach New York fliegen," which implies that it is the Grand Canyon that is doing the flying, in defiance of common sense. In fact, the problem of common-sense knowledge was raised at the dawn of machine translation by the linguist Yehoshua Bar-Hillel, who contrasted "The pen is in the box" with "The box is in the pen."[9] Google Translate returns *Stift* (the writing implement) for both instances of "pen," despite its obvious incorrectness in the latter instance.

These errors reflect the fact that modern machine translation systems like Google Translate are based almost entirely on statistical regularities extracted from parallel text corpora (i.e., texts that have already been translated into multiple languages). Because the writing implement usage of "pen" is vastly more common than the container usage, these systems will fail to appreciate subtle contextual differences that are transparent to humans.

Similar issues arise when computers are asked to answer questions based on natural language. The computer scientist Terry Winograd presented the following two sentences that differ by a single word:[10]

1. The city councilmen refused the demonstrators a permit because they feared violence.
2. The city councilmen refused the demonstrators a permit because they advocated violence.

Who does "they" refer to? Humans intuitively understand that "they" refers to the councilmen in sentence 1 and the demonstrators in sentence 2. Clearly we are using background knowledge about councilmen, demonstrators, permits, and violence to answer this simple question. But building AI systems that can flexibly represent and use such knowledge has proven to be extremely challenging.[11]

As a final example, consider the abilities of a modern image-captioning system.[12] When given the image in Figure 1.1, it returns the caption, "I think it's a person holding a cup." Apparently, the system has implicitly used a heuristic that if it sees a cup and a person in the image, then the image probably shows a person holding a cup. But now consider the image in Figure 1.2, which the same system identifies as "a man holding a laptop." Although the cup is heavily occluded, humans have no trouble recognizing that the person on the left is holding one. And of course the "laptop" is a piece of paper![13]

The lesson from this cursory examination of AI systems is that it is much easier to engineer systems that achieve superhuman performance on specific tasks like Go than it is to engineer systems with human-like common sense. This tells us something very important about the nature of human intelligence: our brains are evolved for "breadth" rather than "depth." We excel at flexibly solving many different problems approximately rather than solving a small number of specific problems precisely. Common sense enables us to make sophisticated inferences on the basis of the most meager data—single sentences or images. And the fact that this ability appears to us so effortless— the very fact that common sense is "common" to the point of being almost invisible—suggests that our brains are optimized for fast, subconscious inference and decision making.

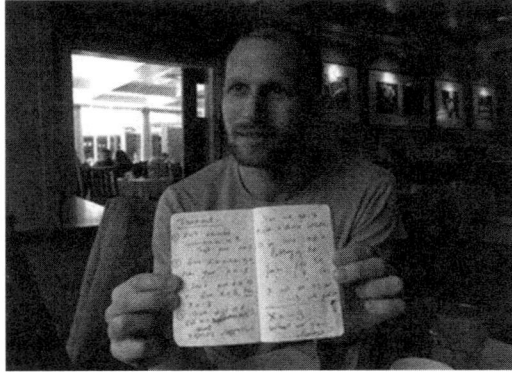

FIGURE 1.1. Image of the author holding a notebook in a restaurant. The image-captioning system believes the image shows "a person holding a cup."

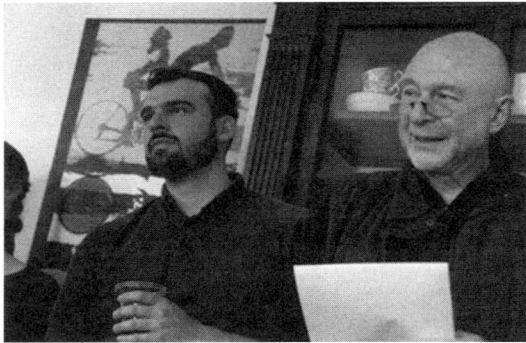

FIGURE 1.2. Image of the author's brother and father. The image-captioning system believes the image shows "a man holding a laptop."

These features of human cognition are shaped by the constraints of the environment in which we live and the biological constraints imposed on our brains. The complexity of our society and technology places a premium on flexibility and scope. We constantly meet new people, visit new places, encounter new objects, and hear new sentences. We are able to generalize broadly from a limited set of experiences with these entities. We have to do all of this with extremely limited energy and memory resources (compared to conventional computers), and under extreme time constraints. To negotiate these demands, our brains make trade-offs and take aggressive shortcuts. This gives rise to errors, but these errors are not haphazard "hacks" or "kluges," as some have argued.[14] They are inevitable consequences of a brain optimized to operate under natural information-processing constraints. The central goal of this book is to develop this argument and show how it reveals the deeper computational logic underlying a range of errors in human cognition.

One might rightfully be concerned that the outcome of this endeavor will be a collection of "just-so" stories—ad hoc justifications of various cognitive oddities.[15] Like Dr. Pangloss in Voltaire's satirical novella *Candide*, we could start from the assumption that "this is the best of all possible worlds" and, given enough explanatory flexibility, explain why all these oddities spring from "the best of all possible minds." However, the goal of this book is not to argue for optimality per se, but rather to show how thinking about optimality can guide us towards a small set of unifying principles for understanding both the successes and failures of cognition. Unlike just-so stories, we will not have bespoke explanations for individual phenomena; the project will be judged successful if the *same* principles can be invoked to explain diverse and superficially distinct phenomena.

I will argue that there are two fundamental principles governing the organization of human intelligence. The first is *inductive bias*: any system (natural or artificial) that makes inferences on the basis of limited data must constrain its hypotheses in some way *before* observing data. For those of you encountering this idea for the first time, it may seem highly unintuitive. Why would we want to constrain our hypotheses before observing data? If the data don't conform to these constraints, won't we be shooting ourselves in the foot? The answer, as I elaborate in the next chapter, is that if all hypotheses are allowable, a huge (possibly infinite) number of hypotheses will be consistent with any given pattern of data. The more agnostic an inferential system is (i.e., the weaker its inductive biases), the more uncertain it will be about the correct hypothesis. Naturally, this gives rise to errors when the inductive biases are wrong. Chapters 2 through 9 are devoted to exploring the implications of this fact, showing the ways in which many different errors that people make are consistent with particular inductive biases. Critically, these are only errors with respect to an *objective* description of reality, to which people do not have direct access.[16] From the *subjective* perspective of an inferential system, the use of inductive biases is not an error at all—it is an indispensable property of a rationally designed inferential system.

The second principle is *approximation bias*: any system (natural or artificial) that makes inferences and decisions with limited resources (time, memory, energy) must make approximations. In particular, optimal inductive inference and planning are intractable for most resource-bounded systems: executing the computations needed to obtain the correct answer requires more time, memory, and energy than is available to these systems. Thus, approximate algorithms are necessary which attain efficiency at the cost of precision. These approximate algorithms give rise to different forms of error, which I explore in Chapters 10 through 12. For example, I show how the need to represent information efficiently leads to distortions in perception, and how the need to calculate probabilities efficiently leads to algorithms that exploit randomness. Again, these are errors with respect to an objective description of reality, whereas they may be optimal from the subjective perspective of the computational system.

2

Rational illusions

There are strange flowers of reason to match each error of the senses.
—LOUIS ARAGON

Shortly after graduating from college, I went on a hiking trip with a friend in New Hampshire. After hiking since dawn, we stopped for a rest on a plateau and gazed at our goal: a mountain peak that loomed in the distance. As we were sitting there, two people came from that direction, and we asked them how long it would take to get to the peak. "About five minutes," they replied. Five minutes?! We stared at the peak in disbelief; it looked distant enough that it would take at least another half hour of hiking. And then, as we were staring, something monstrous appeared on the ascent to the peak: an enormous giant, towering over the trees, surely at least 30 feet tall. After the initial shock, I realized the trick that had been played on my visual system. Since we were above the alpine level, the trees were only about 3 feet tall. But because we were used to seeing much taller trees, our visual system inferred that the peak must be very far away. The giant, of course, was simply another hiker.

The illusion I experienced is illustrative of how size perception is influenced by contextual information. A well-studied example is the Ponzo illusion (Figure 2.1). The two converging lines resemble train tracks that converge into the distance, creating the impression of depth. Consequently, the lower line looks shorter than the upper line, as though it was placed closer to the observer. In fact, both lines are the same length.

While the Ponzo illusion is a contrived example, it relates to the real-world "moon illusion" that has been known since ancient times. At its zenith, the moon is perceived as smaller compared to when it sits at the horizon. The 2nd-century Roman astronomer Ptolemy argued that the moon illusion is caused by the greater apparent distance induced by terrain at the horizon, which seems to fill more space. In support of this argument, the moon illusion can

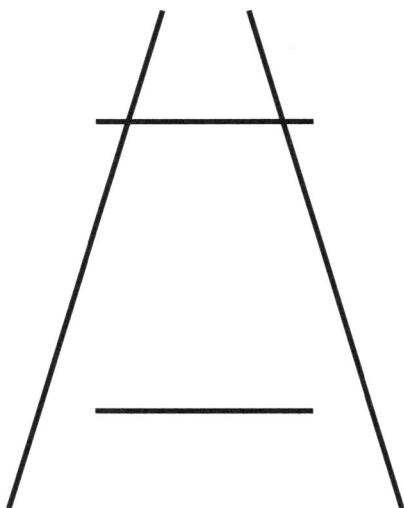

FIGURE 2.1. The Ponzo illusion.

be virtually eliminated by obscuring the terrain (e.g., viewing the moon in complete darkness or through an aperture). In fact, the moon illusion can be reversed by inverting the image so that the terrain appears closer to the zenith moon than the horizon moon.[1]

Contextually induced illusions go far beyond size perception. They appear in the perception of color, location, brightness, speed, weight, and many other properties. The ubiquity of such illusions raises a general question: Why did we not evolve brains that perceive the world as it really is? One answer is that veridical perception is impossible given the limits of our sensory organs.[2] As a consequence, the sensory information that reaches the brain is often highly ambiguous. For example, the three-dimensional world is projected onto a two-dimensional retina. This means that the size and distance of an object are ambiguous: a retinal image could be produced by a small object up close or a large object far away (Figure 2.2). The visual system partially resolves this uncertainty by using contextual information (e.g., perspective cues) and background knowledge (e.g., the canonical sizes of objects).

Illusions are a by-product of ambiguity resolution. The same strategy that aids perception can lead it astray in certain cases (in fact, as I discuss later, it is impossible to devise a strategy that will work well under all circumstances). According to this view, illusions are not bugs, but rather essential design features. If we were to design a robot to optimally perceive the world (within the limits of its sensory receptors), then we would expect it to experience illusions.[3] To unpack this argument, we need to dive deeper into what we mean by an optimally designed system. We can then ask to what extent such optimality principles provide a general explanation for perceptual illusions. Is there a logic of perception?[4]

Perception as inference

Putting our engineering hats on, let us consider how we would endow our robot with a perceptual system. Our robot's input is a retinal image, I, generated by some three-dimensional scene, S, which the robot can't directly

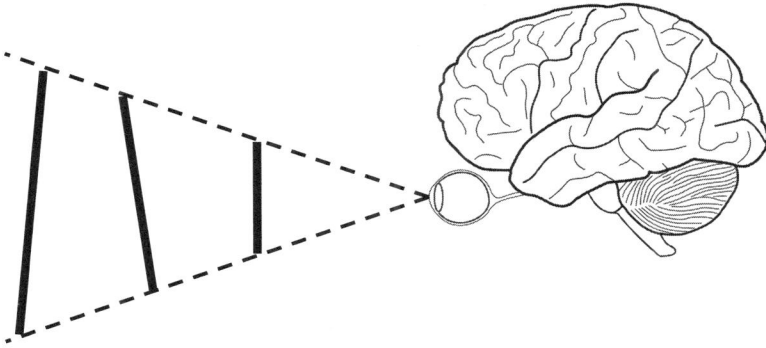

FIGURE 2.2. Sensory information is ambiguous. All of the vertical lines project to the same image on the retina.

observe. As an example, the scene structure could be (partially) specified by the size and distance of a particular object. As already mentioned, the image is typically consistent with many possible scenes. In the Ponzo illusion, for example, a line segment is consistent with an arbitrary size or distance, since we don't have an objective reference point. It's also possible for the image generation process to be "noisy" in the sense that it is influenced by effectively random processes, like whether photons striking the retina cause photoreceptive pigments to change shape, setting in motion the transduction of light into neuronal firing patterns. These different sources of uncertainty can be integrated into a single probability distribution, $P(I|S)$, which expresses the *likelihood* that image I was generated by scene S. Intuitively, the likelihood measures the "fit" between the image and the hypothetical scene.

Note that even if the world were completely deterministic, one can still use probability distributions to express uncertainty (what is sometimes referred to as *epistemic* or *subjective* uncertainty). In our usage, it is better to think of probabilities as degrees of belief rather than descriptions of randomness (the frequencies of repeating events).[5] In conventional usage, I might describe the probability that a coin lands heads by referring to the proportion of heads that I'd expect were I to repeatedly flip the coin. But it's also possible for a weather forecaster to tell you that tomorrow the probability of rain is 70%. Clearly the event "rain tomorrow" can only happen once, so it makes no sense to assign probabilities to a one-time event if we are restricting our usage to frequencies of repeating events. The forecaster is reporting a *belief* about whether or not it will rain tomorrow. This is why Pierre-Simon Laplace, a mathematician who contributed to the early development of probability theory, remarked that probability was "nothing but common sense reduced to calculus."

If our robot has to make a guess about the scene given the retinal image, then one reasonable solution is to report the scene with the highest likelihood (i.e., the scene that is most consistent with the retinal image). This is known as *maximum likelihood estimation*. There are, however, two problems with this solution. First, it neglects contextual information and background knowledge. If you know something about the sizes and distances of nearby objects, or if you know something about the canonical sizes of objects, then you should be able to utilize this information to improve your guess. This leads to the concept of "inductive bias," which I will elaborate later. The second problem is that maximum likelihood estimation neglects subjective uncertainty: if you only report a single guess (a *point estimate*, in statistical parlance), then there's no way to distinguish different levels of confidence in that guess. Suppose you had to bet on your guess; intuitively, you would be willing to bet more if your confidence was higher.

We can remedy the shortcomings of maximum likelihood estimation in two ways. First, we can integrate all of our robot's contextual information and background knowledge into a *prior* probability distribution, $P(S)$. Second, we can allow the robot to report subjective probabilities instead of point estimates. Specifically, it reports the *posterior* probability distribution, $P(S|I)$, the robot's degree of belief that scene S produced image I. One of the most powerful results in probability theory is Bayes' rule, which tells us that the posterior probability is simply the likelihood multiplied by the prior, and normalized so that the probabilities sum to 1:

$$P(S|I) = \frac{P(I|S)P(S)}{\sum_{S'} P(I|S')P(S')}. \tag{2.1}$$

This simple equation has had an enormous impact on theories of the brain and cognition (Box 2.1). We will witness some of its many applications across subsequent chapters in this book.

Box 2.1. The Bayesian brain hypothesis

The idea that the brain represents and manipulates probability distributions might seem exotic at first glance. When I first started studying this question as an undergraduate and told a family friend (a computer science professor) about it, he pointed scornfully at his dog and said, "You think *he* is doing statistics?" But the idea becomes less exotic when we recognize that not only do we routinely report our uncertainty about things, but we can also act in accordance with this uncertainty (e.g., hedging our investments, buying insurance). Bayes' rule has attracted neuroscientists and psychologists because it offers a self-consistent framework for thinking about how uncertainty should be represented, updated, and acted upon.[6] Naturally this doesn't mean that it's correct as a hypothesis about the brain, but it has served as a useful starting point.

How does the brain represent probability distributions? One hypothesis is that populations of neurons implicitly encode distributions.[7] The basic intuition is that downstream neurons receiving signals (spikes) from this population have to reconstruct what information those signals encode, and to do this, they need to take into account the randomness of a neuron's signal-generating process. For example, imagine neurons that fire selectively when particular locations on the retina are stimulated with light. The task for downstream neurons is to reconstruct the stimulated location. Because the firing of neurons is noisy, the best that the downstream neurons can do is assign probabilities to each possible location; these probabilities will be higher to the extent that the noisy neurons selective for those locations are firing more strongly. There are a number of other schemes for representing probabilities with neurons, such as modeling neurons as generating random samples from a distribution[8] (see Chapter 12), or as signaling *prediction errors*[9] (the discrepancy between expected and observed input). These models have been successful at explaining why, for example, the randomness of neural firing seems to track uncertainty (in the case of the random sampling hypothesis), and why expectations about stimuli can sometimes suppress the activity of neurons selective for those stimuli (in the case of the prediction error hypothesis, also known as *predictive coding*).

Although Bayes' rule is conceptually simple, implementing it turns out to be tricky in situations where the denominator cannot be computed exactly (e.g., if there are a very large number of possible scenes). For example, inferring the size and position of even a single object could be computationally intractable. If we discretize the 2 dimensions of object size and 2 dimensions of spatial position into K bins, then to compute the denominator of Bayes' rule exactly would require summing over K^6 possible size-position configurations (a million with $K = 10$). In Chapter 12, we will see a surprisingly effective way to deal with this problem approximately—using random numbers!

Putting aside these computational issues for the time being, suppose now the robot had to act on its beliefs. It chooses an action A and gets rewarded according to $R(S, A)$, where S is the true state of the world (the scene that generated the observed image). The optimal decision rule is to choose the action that maximizes the *expected* reward under the posterior distribution, defined as:

$$\mathbb{E}[R(S, A)] = \sum_{S} P(S|I)R(S, A). \tag{2.2}$$

In other words, the robot should consider the posterior probability of each hypothetical scene, weigh it by the reward associated with acting on that hypothesis, and choose the action that leads to the highest weighted reward when summed across all hypotheses. Putting some technical subtleties aside, the Bayesian decision rule is optimal in the sense that no other decision rule will reliably lead to higher reward.[10]

As an example of how the Bayesian decision rule can be applied, consider the situation in which the robot's action is a guess (point estimate) about the scene, denoted by $\hat{S}$. I offer it X if the estimate is correct, but if it is incorrect, then the robot has to pay me X. Should the robot take this bet? The expected reward in this case is $2P(\hat{S}|I) - 1$, which implies that the robot should only take the bet if the posterior probability of its guess is greater than 0.5 (otherwise the expected reward will be less than or equal to 0). This illustrates how the robot can use its uncertainty to calibrate its betting. This analysis also shows that the robot should (at least for this betting scenario) report the scene with highest probability, also known as the maximum *a posteriori* estimate, if forced to generate a point estimate.

There are several other theoretical arguments about why we would want our robot to be Bayesian. One is the so-called *Dutch book argument*: if the robot did not place its bets according to the Bayesian decision rule, one could create a bet (the Dutch book) that would guarantee the robot a net loss but that nonetheless the robot would accept.[11] Conversely, if the robot follows the Bayesian decision rule, it can guarantee that it won't lose money.[12]

Another theoretical argument is that expressing beliefs as probabilities, and updating them according to Bayes' rule, guarantees that our robot will satisfy an intuitive notion of rationality. Suppose the robot can assign a number, which we'll call a *plausibility*, to each possible hypothesis about the world. Intuitively, logically equivalent hypotheses should have the same plausibility; for example, if two different descriptions refer to the same object, then these two descriptions should be assigned the same plausibility. Small changes in hypotheses should yield small changes in plausibilities, and if a hypothesis is true, it should have a higher plausibility than if it is false. When appropriately formalized, these (and a few other) desiderata lead to the conclusion that plausibilities must be proportional to probabilities, and be updated according to Bayes' rule; any other choice of plausibilities will lead to violations of these criteria for rationality.[13]

Inductive bias

Central to the Bayesian framework is the notion of *inductive bias*: even before our robot has acquired sensory information, it has some prior beliefs about the world. Bayes' rule dictates that these prior beliefs should bias the posterior beliefs, discounting the sensory evidence. This means that a Bayesian robot will make systematic errors. But why would we design a robot that makes systematic errors? The answer is that making inferences about the world is impossible without an inductive bias. Errors are an inevitable consequence of a well-designed inferential robot.

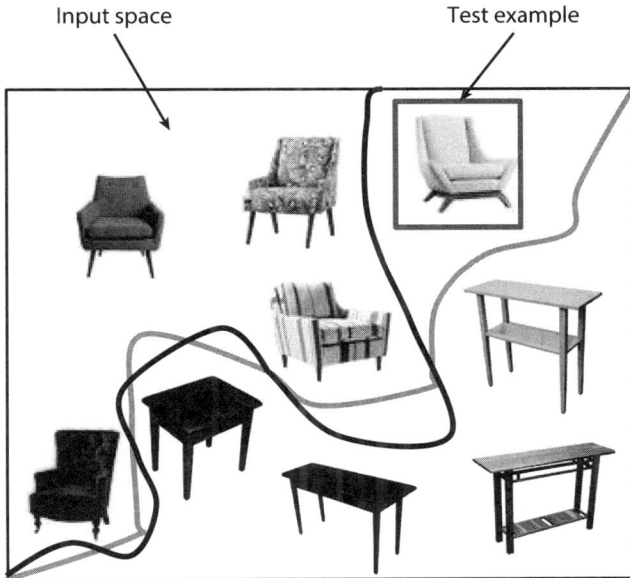

FIGURE 2.3. A chair classifier draws a boundary (possibly complex and non-linear) between chair and not-chair examples in some sensory input space (represented here in 2D). A finite number of examples can be separated by an infinite number of boundaries, assuming the input space is continuous. If none of these boundaries are preferred over others, then the classifier will be unable to classify a test example.

To illustrate this point, I'll borrow an example from the computer scientist Eric Baum.[14] Suppose I build a chair classifier into my robot's visual system. It takes sensory input (e.g., images) and outputs a binary judgment (chair vs. not-chair). Let's suppose I assemble an arbitrarily large (but finite) set of training examples. The classifier is powerful enough to infer a boundary in the input space that separates all the chair examples from all the not-chair examples (Figure 2.3). This means that it can achieve perfect performance on its training set. If the classifier has no inductive bias and the input space is continuous, there exist an infinite number of boundaries that are equally good, in the sense that they all perfectly separate the training examples and thus achieve perfect performance. The implication of this fact is startling: if I give the classifier a new example, it will be unable to determine whether it is a chair or not a chair, no matter how accurate it was on the training set, and no matter how many examples it has collected (short of infinity). There are an infinite number of boundaries that achieve perfect performance and classify the new example as a chair; but there are also an infinite number of boundaries that achieve perfect performance and classify the new example as not-chair. Not having an inductive bias means that the robot has no reason to prefer one of these boundaries over any of the others. Generalization is impossible without an inductive bias!

So inductive bias is necessary, but which inductive bias should we have? It's possible that we could choose a bad inductive bias which would cause us to generalize very poorly, perhaps even worse than random guessing. One way to finesse this problem is to *learn* the inductive bias through repeated experiences with related problems. I will discuss this idea further in the next chapter when we come to the topic of hierarchical learning.

Understanding perceptual illusions

Inductive bias is a key concept for understanding how we perceive the world. Consider, for example, the Kanizsa triangle in Figure 2.4A.[15] One interpretation of this image is that three "Pac-men" are positioned so that their missing sectors form the apices of an equilateral triangle, while three V-junctions are positioned so that their endpoints also form the apices of an equilateral triangle, such that the endpoints intersect the imaginary edges of the triangle formed by the Pac-men. Note that this interpretation does not require us to posit any triangles at all; we simply use triangles to succinctly describe the arrangement of the objects in the image. And yet, we vividly perceive an "illusory surface" formed by a triangle that seems to be implied by the arrangement of the other objects, consistent with an alternative interpretation in which a "camouflaged" equilateral triangle is occluding another triangle flanked by three black circles.

Why do our brains prefer the occlusion interpretation? Arguably because it would require a highly improbable coincidence to arrange the Pac-men and V-junctions in just the right way, whereas the occlusion of one surface by another is quite common. The occlusion interpretation should be more generalizable across different viewing conditions and slight perturbations of the scene. For example, rotating the occluding surface should leave the effect intact (Figure 2.4B). Even after rotating the bottom two Pac-men so that their "mouths" point slightly upwards, one continues to see an illusory surface, as though the triangle was folded at the corners (Figure 2.4C).[16] Slightly rotating the Pac-men and V-junctions in random directions, as though one bumped a table overlaid with a fragile arrangement of cutouts, diminishes the effect (Figure 2.4D), but even in this case one can discern a surface that has been folded and partially cut.

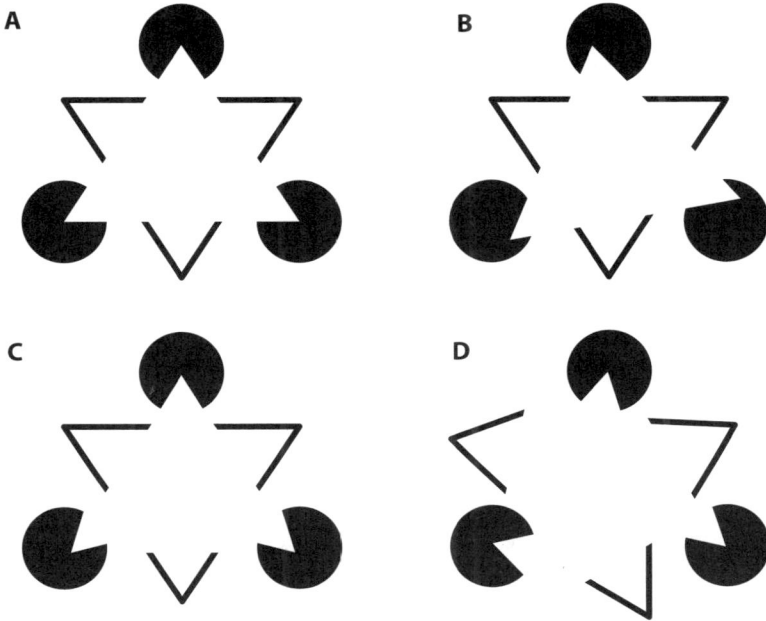

FIGURE 2.4. (A) The Kanizsa triangle. (B) Rotating the invisible occluding surface leaves the illusion intact. (C) If the bottom two Pac-men are rotated slightly, one sees a bent occluding surface. (D) If the image is a suspicious coincidence of Pac-men and V-junctions, then jittering the arrangement largely destroys the illusory surface.

These examples illustrate how our visual system has strong inductive biases about the structure of the world. Though many different scene inter-pretations are consistent with our sensory inputs, we strongly prefer certain interpretations over others.

How far can we go with this framework? As we will see later, not all errors are naturally derived from inductive biases. Before we get to that point, it will be instructive to go through a few more examples in greater detail. These examples were chosen to illustrate two fundamental principles that arise naturally from the rules of probability:

1. *Explaining away*: When several hypotheses can potentially explain the same observations, additional evidence for one of the hypotheses reduces belief in the other hypotheses.[17]
2. *Integration*: When several sources of information about a hypothesis are available, each source influences beliefs about the hypothesis in

proportion to the source's precision (the accuracy of the information it provides).

Explaining away and lightness illusions

When a surface is illuminated by a light source, a proportion of it (the *reflectance*) bounces off the surface. Some of this light (the *luminance*) reaches our retina, activating photoreceptive neurons. One of the problems facing the visual system is to reconstruct the reflectance from retinal measurements of luminance. *Lightness* is the subjective estimate of reflectance. Whereas reflectance and luminance are physically measurable quantities, lightness is a perceptual quantity that can only be measured by self-report.

Reflectance is ambiguous, because a particular luminance could be produced by a highly reflective (shiny) surface under low levels of illumination (dim light), or by a less reflective surface under high levels of illumination (note the analogy to the size-distance ambiguity discussed earlier in this chapter). The brain uses a number of visual cues to resolve this ambiguity. A classic example of this ambiguity resolution is the Craik-O'Brien-Cornsweet illusion (Figure 2.5).[18] Two adjacent squares have identical surfaces, whose reflectance (and hence luminance) decreases gradually from left to right. Despite the fact that they are identical, the left square is perceived as slightly darker than the right square.

One explanation for this error is that the brain has a strong inductive bias to assume that a light source will not uniformly illuminate a surface unless it is placed directly above the object (an improbable coincidence). More typically, objects and light sources are not aligned with one another, and therefore the light hitting the surface of an object will manifest as a gradient of illumination. This gradient "explains away" the reflectance gradient; the visual system can explain the luminance gradient in terms of differences in illumination, and since it would be improbable for there to be both reflectance and illumination gradients, the brain prefers only one of these explanations.

If other cues are available, the brain may resolve the ambiguity differently. Figure 2.6 shows a three-dimensional version of the Craik-O'Brien-Cornsweet illusion using bricks that appear to be painted different shades of gray. In fact, both bricks have identical reflectance gradients, as in the two-dimensional version of the illusion. When we apply the same reflectance gradients to two cylinders, the illusion is attenuated, because now object shape (surface curvature) offers another plausible explanation of the luminance gradient.

The same line of reasoning can explain why a surface appears darker when placed on a light background: the inferred surface reflectance is lower if the

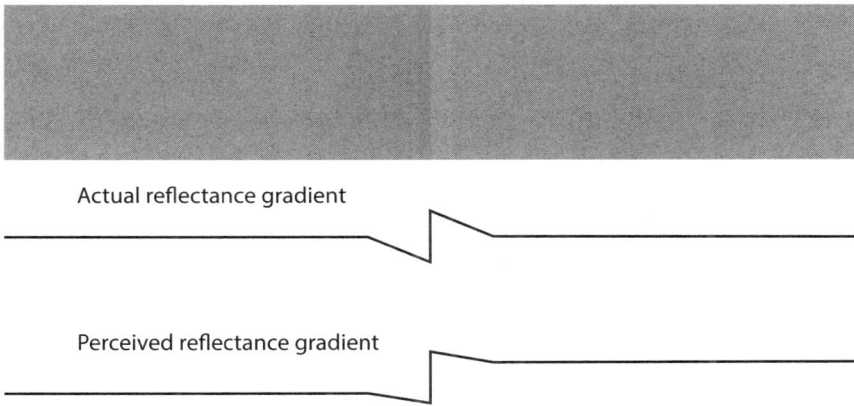

Actual reflectance gradient

Perceived reflectance gradient

FIGURE 2.5. The Craik-O'Brien-Cornsweet illusion. *Source:* Wikipedia.

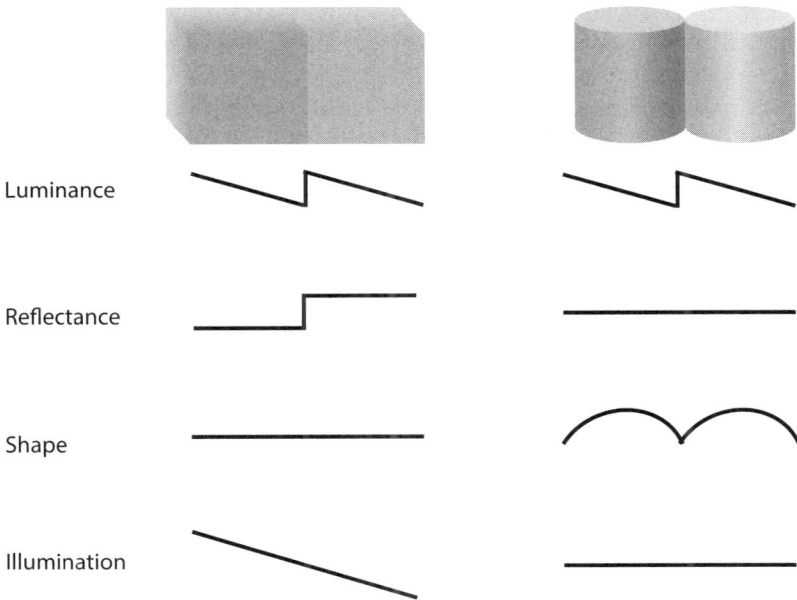

Luminance

Reflectance

Shape

Illumination

FIGURE 2.6. The brick on the left appears darker than the brick on the right, as though they were painted with different shades of gray, but in fact both bricks have the same reflectance. This illusion is attenuated for the cylinders, because the shape (surface curvature) "explains away" the luminance gradient. Below each stimulus is the scene interpretation favored by the visual system. Adapted from Adelson (2000) and Knill and Kersten (1991). Courtesy of Springer Nature.

FIGURE 2.7. Koffka rings.

inferred scene illumination is higher. Intuitively, shining a light on an object will not make the object appear to have a lighter shade, even though the luminance reaching your retina is greater. Illumination explains away reflectance. Nonetheless, this explaining away can be overridden by other cues. For example, the Koffka rings in Figure 2.7 illustrate how surface cues can produce the appearance of uniform reflectance. When two half-rings are connected to form a ring, the contrast illusion disappears, because the scene is more plausibly explained by a single surface occluding two backgrounds with different reflectances, rather than two surfaces with different reflectances occluding a single background with uniform reflectance.

The idea that the brain seeks "explanations" of its visual inputs, and evaluates the quality of explanations using Bayes' rule, can be contrasted with "low-level" accounts of illusions, which attribute them to various kinds of image-filtering operations implemented by the brain. For example, some lightness illusions have been attributed to neurons that are excited in response to luminance in a particular region of visual space and are suppressed in response to luminance in neighboring regions. These "center-surround" neurons collectively have the effect of enhancing edges in the neural representation of an image, which in some cases produces illusory lightness. In principle, all illusions could be accounted for by positing appropriate neural filtering operations. The basic problem facing such accounts is that they cannot do justice to the bewildering flexibility of the visual system—it would require a baroque collection of filtering operations akin to Ptolemaic epicycles. How, for example, would these operations "know" about the three-dimensional shape of objects and adjust their filter parameters in just the right way?

The Bayesian framework lifts this problem to a higher level of abstraction by placing the explanatory burden on the internal models assumed by the brain. As long as these internal models are sufficiently flexible (e.g., different luminance patterns can be explained by different combinations of illumination, shape, and reflectance), then Bayes' rule (and, more generally, the probability calculus) offers a coherent mechanism for reasoning about them. This does not deny the existence of neural filtering operations, but rather constrains what kinds of operations the brain would need to implement.

Integration and ventriloquism

Nearly everybody is familiar with the ventriloquist act: a puppeteer coordinates her voice and the puppet's movement in such a way that the puppet appears to be talking. We really feel as though the voice is emanating from the puppet. How is this feat accomplished?

From the observer's perspective, we can formalize this scenario as a problem of multi-sensory cue combination. The observer needs to integrate auditory and visual cues about the spatial location of the speaker. In a laboratory setting, this can be studied by presenting human subjects with auditory and visual cues simultaneously, and then asking them to indicate the location of the audiovisual source on a one-dimensional axis (e.g., horizontal positions on a computer screen) or to judge whether the source location was to the left or right of some preceding reference event. The standard finding is that vision "captures" sound: localization of the source is systematically biased towards the visual location.[19]

A simple mathematical model can explain this phenomenon (Figure 2.8). Assume for simplicity that the prior distribution over location is uniform, so that no location is preferred a priori. Then Bayes' rule says that the posterior over location S given sensory information I is proportional to the likelihood:

$$P(S|I) \propto P(I|S) = P(I_A|S)P(I_V|S). \qquad (2.3)$$

The sensory information consists of two parts—auditory (I_A) and visual (I_V)—each corresponding to a location sampled from distributions centered on the true (but unknown) location S. The widths of these distributions depend on the precision of each sensory modality, with broader distributions for less precise modalities. At a physical level, precision derives from many different sources (e.g., the structure of the retina and cochlea), but we can operationally define precision as the average accuracy of spatial localization when a cue from a single modality is presented. Humans are less accurate at localizing auditory cues compared to visual cues. Consequently, the posterior distribution is biased towards the mean of the visual location distribution—a simple laboratory analog of the ventriloquist illusion.

Estimated
location

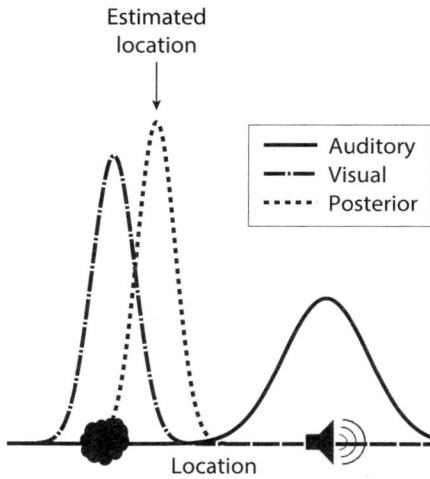

The Bayesian model makes another striking prediction: the ventriloquist illusion can be reversed! Specifically, if the precision of the location information provided by the visual cue is sufficiently degraded (e.g., by blurring or enlargement), then the auditory cue will have higher relative precision, causing sound to capture vision.[20]

The story does not end there, however. It turns out that multisensory integration (and hence the ventriloquist illusion) breaks down when the discrepancy between auditory and visual information is large.[21] Thus, the simple model in which auditory and visual cues are obligatorily integrated seems to fail under these conditions. One explanation is that people are making inferences about the causal structure of the cues.[22] Integration occurs when people infer that a single object (hidden cause) produces both auditory and visual cues. This single-cause hypothesis becomes increasingly improbable when the location information provided by the two cues is highly discrepant. Instead, a multiple-cause hypothesis becomes more probable, according to which the two cues are generated by different objects. Indeed, when asked directly, people are more likely to report that a single object generated both cues under low-discrepancy conditions.[23]

FIGURE 2.8. Ventriloquism as optimal cue combination. Auditory and visual signals are sampled from observation distributions and then combined via Bayes' rule to produce a posterior distribution. The location with the highest posterior probability is usually taken as the subjective estimate of the source object's location. Because visual precision is higher than auditory precision, the subjective estimate is biased towards the mean of the visual distribution.

Cognitive illusions

A key idea of this chapter is that many perceptual illusions are fundamentally cognitive, in the sense that they draw upon high-level knowledge. At the same time, high-level cognition is itself susceptible to many illusions. One view of cognitive illusions attributes them to heuristics that, while typically useful, generate systematic errors.[24] These heuristics are analogous to the image-filtering operations discussed above in the context of lightness illusions, and they run into some of the same explanatory problems. They can account for specific illusions, but they don't offer a coherent account of cognitive flexibility: What allows us to adapt to the wide range of circumstances that we regularly face? Just as it did for perceptual illusions, the Bayesian framework places the explanatory burden on internal models combined with probabilistic reasoning. Cognitive flexibility derives from the richness of the mind's internal models and the versatility of Bayes' rule. To illustrate this point, I will consider a few examples here that parallel the perceptual examples in the previous section. We will encounter many other examples in subsequent chapters.

Explaining away and the fundamental attribution error

Most of human behavior is a function of personal disposition (e.g., how nice a person is) and situational factors (e.g., cultural norms). In 2004, journalists uncovered evidence of prisoner abuse at Abu Ghraib prison in Iraq, leading to the dishonorable discharge of several soldiers. When interviewed on CBS, Brigadier General Mark Kimmitt made the argument that this was the action of a few bad apples:

> So what would I tell the people of Iraq? This is wrong. This is reprehensible. But this is not representative of the 150,000 soldiers that are over here . . . I'd say the same thing to the American people . . . Don't judge your army based on the actions of a few.[25]

In other words, Kimmitt is making a dispositional inference about the soldiers, discounting the situational factors that may have influenced their behavior. The focus on the responsibility of individual actors, rather than situational factors, is characteristic of human social judgment—so characteristic that this tendency has been designated the *fundamental attribution error*.[26] The reason it is called an error is because people seem to inadequately take into account the power of situational factors, even when they are highly relevant. Inspections of Abu Ghraib prison by the International Committee of the Red Cross led to the conclusion that the abuses were not isolated acts, but rather part of a "pattern and broad system."[27]

This conclusion echoes Hannah Arendt's famous "banality of evil" argument that Adolf Eichmann's crimes during the Holocaust were not the idiosyncratic acts of an unusually evil individual.[28] The Nazis had created a legal system that justified and normalized acts considered crimes by people outside the Nazi system. Eichmann was just doing his job. Arendt's argument sparked a huge amount of controversy, in part because it seems to contradict our strong inclination to assign responsibility to individuals rather than to situations.[29]

Stanley Milgram's studies of obedience to authority reinforce this observation. Milgram took ostensibly normal people off the street of New Haven and asked them to act as a "teacher," delivering electric shocks to another individual (the "learner") whenever the learner gave an incorrect response to a question.[30] The shocks increased in voltage for each incorrect answer. In reality, there were no shocks, and the learner was a confederate pretending to be shocked, but from the teacher's perspective everything was quite real. Despite their visible discomfort, most teachers, when commanded by the experimenter, continued delivering shocks with higher and higher voltages, as the learner's expressions of pain and protest gave way to screams and ultimately complete silence. I recall watching videos of these experiments as a high school student and being astonished that "normal" people could, under sufficient pressure, commit what looked like murder. My astonishment derived from the fundamental attribution error: my mind resisted the inevitable conclusion that these people were not dispositionally "bad." Like Eichmann, they were just doing their job.[31]

Psychologists have investigated the fundamental attribution error more directly by asking people to make dispositional inferences about actors after receiving various kinds of situational information. In one classic study, university students read an essay (supposedly written by a classmate), which gave a favorable or unfavorable opinion about Fidel Castro.[32] In the free-choice condition, the students were told that their classmate was free to write a pro or con essay, whereas in the forced-choice condition, the students were told that their classmate was instructed to write a pro or con essay. Not surprisingly, students in the free-choice condition judged the attitude of their classmate towards Castro to be consistent with the opinion expressed in the essay. Critically, this was still true (albeit more weakly) for the students in the forced-choice condition. In other words, they failed to completely explain away the situational constraints in forming a dispositional inference.

The normative question here is whether people discount enough relative to a rational standard of inference. If one assumes that instructions deterministically control behavior, then the answer is no: discounting should be complete, but empirically it is not. We know, however, that instructions are rarely so potent. When asked about the probability that a classmate with an anti-Castro attitude would write a pro-Castro essay in the forced-choice condition, students judged this probability to be 0.85. Thus, instructions are not considered sufficient to produce behavior.[33] Indeed, 35% of people in the

Milgram experiment refused to deliver the highest-voltage shock despite the experimenter declaring that they *must* go on.

Once we take into account the fact that situational factors do not deterministically control behavior, a Bayesian analysis naturally produces the observed behavior (partial discounting).[34] The Bayesian analysis also makes a number of predictions that are consistent with other studies. Discounting can be amplified by explicitly telling the students that the instructions are sufficient to compel behavior,[35] or by telling them that people with pro-Castro attitudes were selected to receive anti-Castro instructions.[36]

We can draw an analogy between the Bayesian analysis of dispositional inference and the analysis of lightness illusions given above. In both cases, multiple variables interact to produce observations, rendering the interpretation of observations ambiguous. This ambiguity can be partially resolved by additional information, which is interpreted in accordance with Bayes' rule applied to a particular internal model of the environment. When additional information favors one hypothesis, this causes alternative hypotheses to be explained away.

Integration and metacognitive errors

Metacognition—thinking about thinking—is sometimes held up as a hallmark of human intelligence (though there is substantial evidence that some animals can also engage in metacognition). We not only can report our memories, feelings, sensations, and decisions—we can report judgments *about* those memories, feelings, sensations, and decisions. For example, when we make a choice, we can assess our confidence that the choice was correct (i.e., that no other choice will lead to a better outcome). A basic question about such confidence judgments is whether they are well-calibrated: Does the objective likelihood of being correct (accuracy) correspond with our subjective confidence ratings?

While in general accuracy and confidence are correlated, they are not in perfect correspondence. In particular, people typically are underconfident on easy tasks and overconfident on hard tasks, a phenomenon known as the *hard-easy effect*.[37] Clearly this behavior is suboptimal, but is it irrational? To make the case for rationality, we would need to demonstrate that people are doing the best they can given their limited access to information.

In tasks that measure confidence, people do not typically have direct access to their accuracy; this has to be estimated from experience. People start with some prior belief about accuracy, and then get feedback based on their task performance. If they are rational, they will update their prior belief using Bayes' rule to form a posterior belief.[38] First, consider the case where a person is more accurate than expected under the prior (the task was unexpectedly easy). Their posterior expectation will then be somewhere in between the prior and the feedback signal, which means that the person will underestimate accuracy on relatively easy tasks. Next, consider the case where the person is less accurate than expected. The posterior expectation will again be somewhere in between the prior and the feedback signal, which means that the person will overestimate accuracy on relatively hard tasks. In summary, the hard-easy effect is consistent with rational belief updating.

One implication of this analysis is that the hard-easy effect should be modulated by the amount of evidence. When people have access to more evidence, confidence judgments should be pulled less strongly towards the prior, causing less underestimation on easy tasks and less overestimation on hard tasks. Consistent with this prediction, the hard-easy effect is stronger when making judgments about the accuracy of other people, about whom evidence is less available.[39]

We can draw an analogy between the Bayesian analysis of confidence and the analysis of ventriloquism given above. In both cases, multiple variables need to be integrated to form a belief about a hidden variable. Bayes' rule dictates how to rationally combine these variables based on the precision of information that they provide.

Summary

Perceptual errors are defined with respect to an objective ground truth inaccessible to our brains. Rationality is a claim about what our brains do with the information they get from the senses. Thus, it is perfectly possible to make rational errors. Indeed, the concept of inductive bias implies that a rational information-processing system will always make errors. By examining the specific errors that people make, we can reverse engineer the structure and origins of their inductive biases, as I explore in the next chapter.

3

Structure and origins
of inductive bias

A mind perpetually open, will be a mind perpetually vacant.

—BERTRAND RUSSELL

In the last chapter, I stressed the importance of inductive biases using specific examples. Now I'll sketch the bigger picture: What kinds of inductive biases do people have, and how general are they?

Causality

Some inductive biases are sufficiently pervasive that we can consider them domain-general. One such inductive bias is causality: we perceive events as contingent on other events, not merely correlated with them. For example, when a moving object touches a stationary object and the stationary object then begins to move, we perceive the moving object as causing the movement of the stationary object.

The psychologist Albert Michotte showed that this impression of causality can be generated by very sparse visual information.[1] An object (as shown in Figure 3.1) collides with another object, and immediately the second object begins moving while the first object remains stationary. For many observers, this display produces the strong perception that the first circle *caused* the second circle to move. This perception is diminished if a temporal or spatial gap is interposed between the two objects' motions, indicating that we expect causes to act immediately upon physical contact. Michotte's findings demonstrate that causality can be apprehended from extremely sparse information; the observers know virtually nothing about the structure of the environment in which these perceptual events are occurring.

Launching

Launching with a temporal gap

Launching with a spatial gap

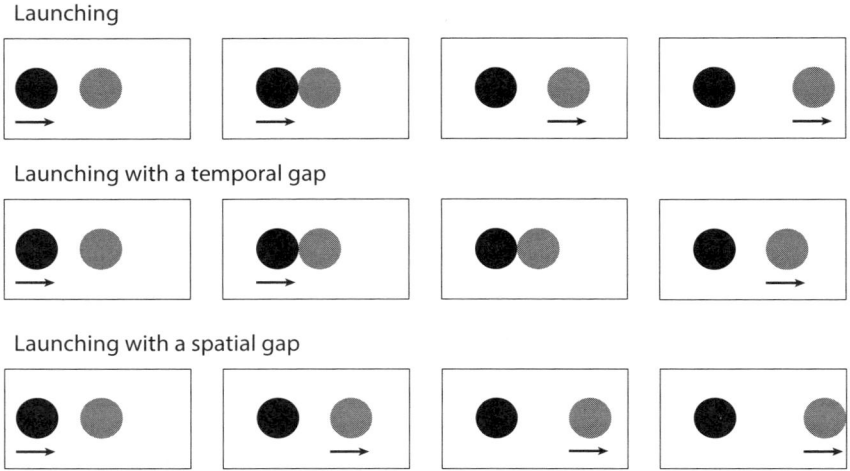

FIGURE 3.1. Perceiving physical causality: the object on the left appears to cause the motion of the object on the right when they collide, unless there is a temporal or spatial gap between the two objects' motions.

Michotte's experiments capture a notion of *physical causality*. When we think about other agents (people, animals, even some robots), we also invoke a notion of *mental causality*: an agent's actions are contingent on having certain beliefs and desires. If a person collides into another person, it's usually because they wanted to, or because they didn't see the other person.

Around the same time that Michotte was showing people colliding objects, the psychologists Fritz Heider and Marianne Simmel were showing people animations of simple geometric objects (Figure 3.2), which people perceived as agents, despite no contextual information or recognizably agent-like visual features.[2] With both the Heider-Simmel and Michotte animations, the important point is that people need very little data to make strong inferences about physical and mental causality—precisely the signature of an inductive bias.

The power of causality is that it can work as a kind of universal "glue" for many different kinds of events (physical, mental, conceptual, etc.), parsimoniously explaining how events fit together. As we've already seen, domain-general causal reasoning draws upon domain-specific knowledge about physics (e.g., mass, force, acceleration) and agency (e.g., beliefs, desires, plans). Some have argued that this knowledge is organized, starting in childhood, into "intuitive theories" that resemble scientific theories: causal models that explain why things happen, what would happen if we intervened in certain ways, and what would have happened in counterfactual situations.[3]

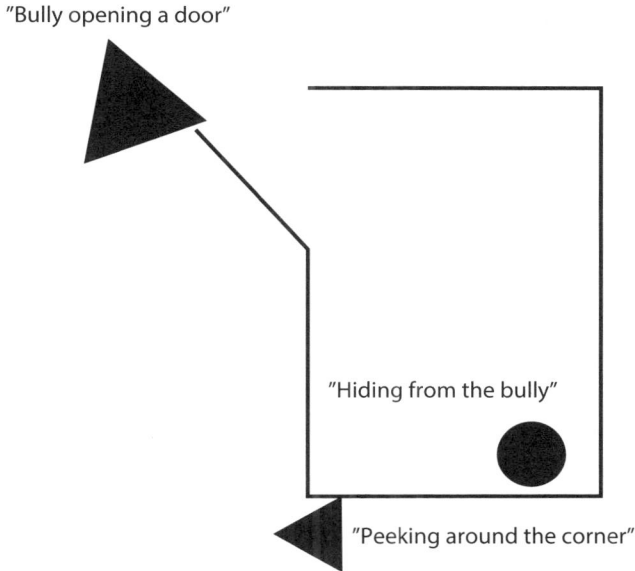

FIGURE 3.2. Perceiving mental causality: simple geometric objects appear to act based on beliefs and desires.

As with all inductive biases, causal inductive biases can lead to errors: people sometimes see causality where none exists, such as the medieval belief that blood-letting cured illnesses, or the belief that vaccines cause autism. Although these beliefs are factually incorrect, they are not necessarily irrational if we assume that belief systems are structured around inductive biases that make these inferences plausible (a point I will discuss further in Chapter 6).

Compositionality

Another domain-general inductive bias is the propensity to see complex things as composed of simpler things. Society is composed of individuals, objects are composed of parts, sentences are composed of phrases, plans are composed of sub-plans, spatial regions are composed of sub-regions, and so on. This "divide and conquer" strategy allows us to represent a large, possibly infinite, number of concepts with a small number of building blocks.

Wilhelm von Humboldt invoked compositionality when he described language as "the infinite use of finite means." A classic example is the English past tense: children learn that the past-tense form of most verbs can be produced by concatenating the present-tense form (e.g., *cook*) with the suffix "-ed"

Geons   Objects

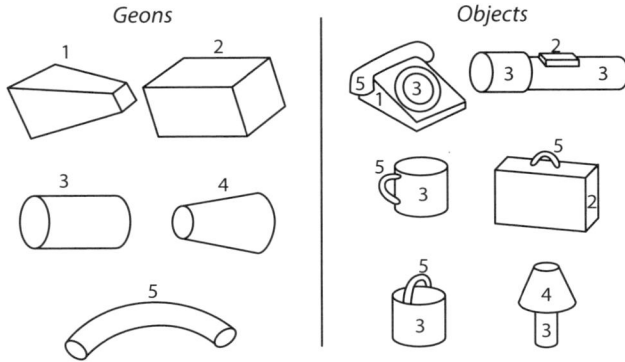

FIGURE 3.3. Part-based decomposition of objects into "geons."
Adapted from Biederman (1995).

(e.g., *cooked*). Verbs obeying this rule are referred to as *regular*; a small proportion of verbs are *irregular*, obeying idiosyncratic rules (e.g., the past tense of *drive* is *drove*). This has led to the hypothesis that our brains have separate representations of compositional linguistic rules and exceptions to those rules.[4] Consistent with this hypothesis, the regular past tense can be generalized to novel verbs, whereas the irregular past tense is typically not generalized as broadly unless the novel verb is phonologically similar to familiar irregular verbs.[5] For example, people will generalize the irregular *swing-swung* to *spling-splung*, but not to *nist-nust*.

One of the most recognizable signatures of compositional generalization is the tendency to *over-regularize* the English past tense: as children acquire the regular past tense, they begin to over-apply it to irregular verbs (e.g., *run-runned*). This is another example of how inductive biases can produce systematic errors. Nonetheless, the rate of over-regularization is still quite low (estimates range between 2.5% and 10%).[6] Counterbalancing these errors is the enormous benefit for language learning: compositional rules allow children to generalize far beyond their meager linguistic input (what Chomsky called "the poverty of the stimulus").[7]

The inductive power of compositionality extends beyond language to other domains of cognition. I'll illustrate with a few examples from visual perception.

The psychologist Irving Biederman recognized that many objects, particularly man-made objects, can be decomposed into a small number of geometric primitives that he dubbed "geons" (Figure 3.3).[8] The geon theory was motivated by the observation, already touched on earlier, that certain properties of images are *non-accidental*: they are highly unlikely to arise from the accidental

Complete Component Deletion Midsegment Deletion

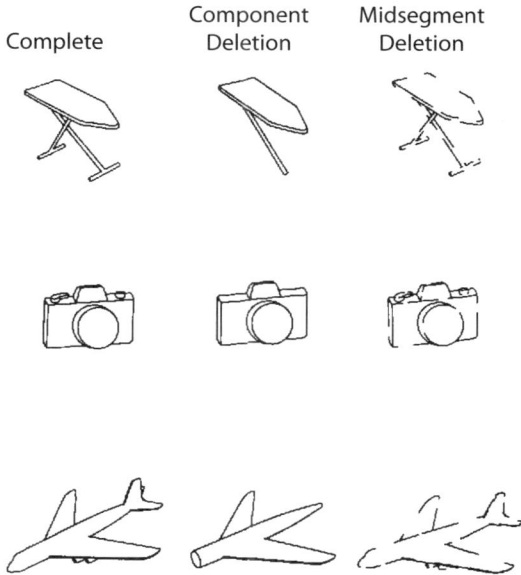

FIGURE 3.4. Stimuli used in experiments reported by Biederman (1987).

alignment of viewpoint and object features.[9] For example, a straight line in an image could theoretically be produced by the curved edge of an object viewed from a particular angle, but in the absence of additional evidence supporting such a hypothesis, our visual system interprets the straight line as arising from the straight edge of an object. Biederman then asked what kind of part-based decomposition of three-dimensional objects would admit rapid identification on the basis of non-accidental properties of two-dimensional images. By applying a set of transformations (rescaling, bending, and reflecting) to regions of a cylinder, Biederman constructed a set of 36 geon primitives that could be attached together to form a large number of recognizable objects.

The geon theory stresses two claims: (1) parts can be rapidly identified from non-accidental properties, and (2) the part-based decomposition is used to recognize object categories. Thus, removing non-accidental properties from images should impair object recognition. Biederman tested this prediction in a number of clever experiments. For example, he removed equivalent proportions of image contours either at midsegments or as whole components (Figure 3.4), finding that component removal reduced recognition performance much more strongly than midsegment removal when the images were presented very briefly. Another striking example is the effect

Non-recoverable   Recoverable

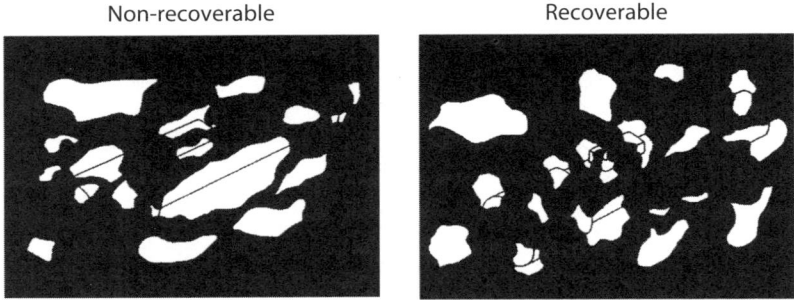

FIGURE 3.5. Occluding non-accidental properties (left) makes object recognition very difficult, whereas occluding other regions (right) does not. Adapted from Biederman (1987).

of occluding non-accidental properties (Figure 3.5), which strongly degrades object recognition.

Geons are not the only way to decompose images. Another approach is to take seriously the idea that images are *generated* via some physical process. For example, the cognitive scientist Brenden Lake and his collaborators collected a large number of handwritten characters from 50 different writing systems.[10] These characters were generated by composing together a set of canonical strokes drawn from a motor vocabulary. Lake and his collaborators argued that people recognize exemplars of particular characters, as well as generate new exemplars, using a mental representation of this generative process (Figure 3.6), learned through observation of many different characters. Once this mental representation is learned, new character concepts can be assimilated with as few as a single exemplar.

So far we have been talking about static images, but real-world scenes are filled with motion. Our visual systems use a divide-and-conquer strategy here too, parsing complex dynamic scenes into a hierarchy of relative motion components (Figure 3.7).[11] For example, the motion of Spiderman's hand can be decomposed into the motion of the hand relative to the arm, the arm relative to the body, the body relative to the train, and the train relative to the background.

A classic demonstration of relative motion was devised by Karl Duncker, and is now known as the Duncker wheel.[12] If you place a light on the rim of a wheel and then roll the wheel in the dark, the light will look like it's bouncing in an arc (Figure 3.8), known technically as *cycloidal* motion. If you add another light to the hub, then the outer light's motion is now interpreted as rotation around a horizontally translating reference frame.

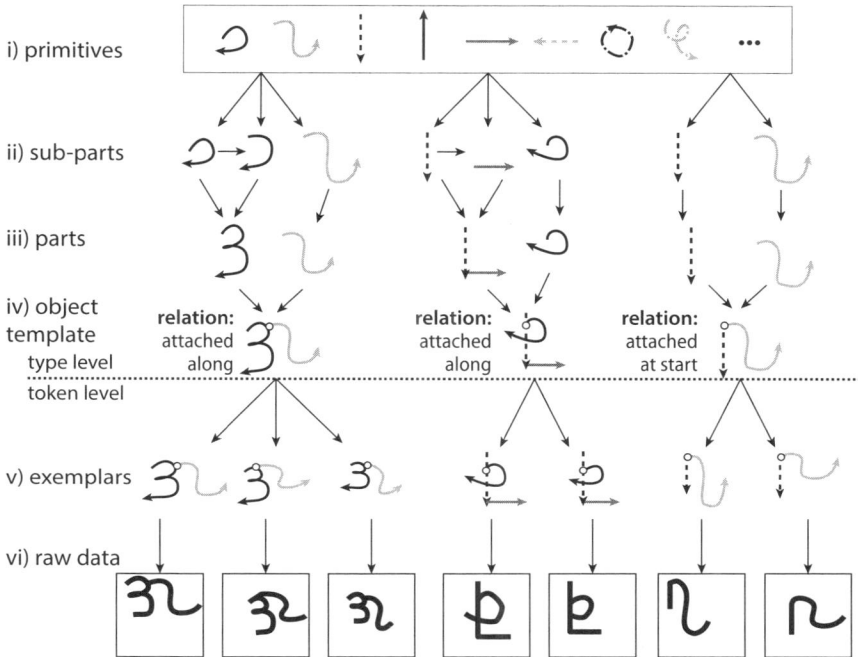

FIGURE 3.6. Decomposition of handwritten characters into strokes. From Lake et al. (2015). Copyright © 2015 by The American Association for the Advancement of Science. Reprinted with permission from AAAS.

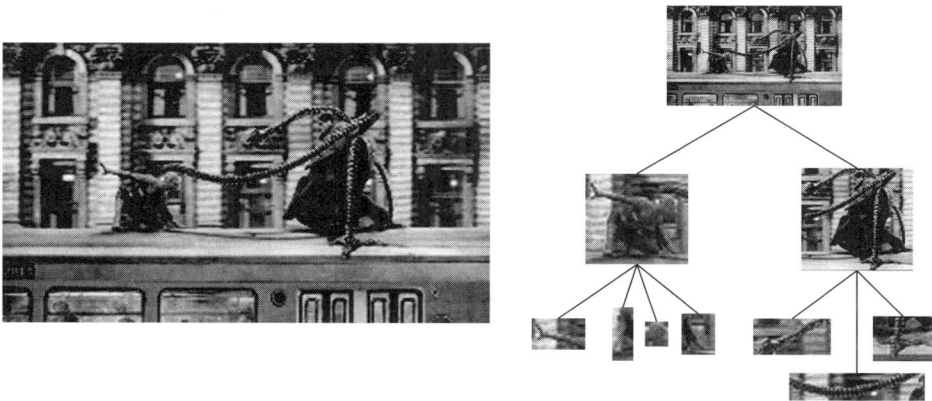

FIGURE 3.7. A complex dynamic scene decomposed into a hierarchy of relative motion components. From Gershman et al. (2016). Copyright © 2016 by Elsevier. Reprinted by permission.

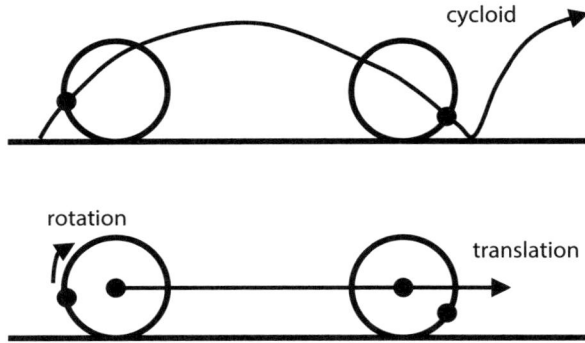

FIGURE 3.8. The Duncker wheel.

Relative motion decomposition can induce velocity illusions. For example, a moving object will appear faster or slower depending on the speed of objects around it,[13] and objects that are moving in random directions can be perceived as moving coherently in one direction, depending on the motion direction of nearby objects.[14]

Objects

You may have noticed that I repeatedly refer to things called "objects"—one object causes another object to move, parts are composed together to form objects, and so on—taking for granted in each case that our mental representations of the world are populated by objects. The fact that you may not have noticed is telling: objects are so pervasive in our thinking that they can slip into our scientific theories without generating any controversy. But it is quite possible to build robots that do not see the world in terms of objects. In fact, many computer vision systems work without any explicit representation of objects, as do many models of visual processing in the human brain.[15] These models are missing something important about the nature of perception.

Representing the inanimate world in terms of objects is a strong inductive bias: our understanding of causality and compositionality is not defined with respect to features like color, shape, size, pixels, but rather with respect to objects. It excludes pixels from causing other pixels to move, and it excludes composing objects out of pixels. An object-oriented inductive bias allows us to learn rapidly and generalize broadly, to the extent that properties of the world adhere to objects.

One reason objects are useful as an inductive bias is that, unlike low-level properties, they change slowly across space and time, and therefore allow us to generalize our knowledge more persistently. The image on your retina is in constant flux, due not only to motion in the world but also motion of your eyes and head. Objects in the world tend to move slowly and continuously, a fact that your brain exploits. For example, it is much easier for us to track objects that move in a spatiotemporally continuous manner, compared to if they wink in and out of existence.[16] Strikingly, spatiotemporal continuity can override other cues to objecthood; if an object passes behind an occluder and reappears on the other side with a different shape or color, people still perceive it as a single object, provided it appears in the time and location dictated by a constant velocity (a phenomenon known as the *tunnel effect*).[17] Conversely, an object that maintains its features but reappears in the wrong location or at the wrong time is perceived as a different object.

Organizing the world in terms of objects can produce powerful visual illusions. For example, if an object appears in one location, disappears, and then quickly reappears at another location, we tend to perceive the object as moving along a smooth trajectory from the first location to the second location, a phenomenon known as *apparent motion*.[18] This effect disappears if the spatial and temporal gaps are too large. As with the tunnel effect, apparent motion can override other perceptual features,[19] again demonstrating the primacy of spatiotemporal cues in object perception.

The origin of inductive biases

Where do inductive biases come from? All of the inductive biases discussed above appear in some proto-form early in development, indicating that they might be built in by evolution. For example, some evidence suggests that newborn infants perceive partly occluded objects as connected units.[20] Some authors have speculated that evolution equips us with inductive biases by encoding the instructions for brain wiring in the genome.[21] Intriguingly, it is possible to construct artificial neural network architectures that perform surprisingly well *in the absence of learning*. For example, one can search over the space of network topologies (the connection patterns between neurons) with random connection strengths to find particular topologies that do better than others in a manner that is invariant to the particular choices of connection strengths.[22] This suggests that structural constraints are computationally plausible means of implementing some inductive biases.

Inductive biases also develop in important ways over the course of childhood. Young infants look longer at displays in which objects are not spatiotemporally continuous, which is typically taken to indicate that their expectations were violated, while sensitivity to gravity and inertia develop later.[23] The perception of causality in collision events develops very early, and is perhaps even innate,[24] but a fully functional conception of causality develops later. For example, even two-year-olds do not spontaneously intervene in causal systems to bring about an effect.[25] Young children also appear to have a limited ability to understand compositionality.[26]

One way to understand these developmental patterns is via learning: we can apply the same principles of Bayesian inference over hypotheses to the inductive biases themselves.[27] Formally, this is known as a *hierarchical* Bayesian model, because it involves defining a hierarchy of priors (i.e., priors over priors). This framework naturally gives rise to *learning-to-learn* phenomena, whereby experience informs not only the task at hand but the entire *distribution* of tasks that a person may encounter.[28]

Concretely, consider the problem of word learning: a child simultaneously receives perceptual input and spoken words, and must figure out what in the perceptual input the words refer to. This problem is highly ambiguous, because the number of possible referents is virtually unlimited. However, children do not consider all of these possibilities; for example, they assume that object names refer to whole objects, not to parts of objects or their properties (the *whole object bias*[29]), and they assume that object names generalize to new objects on the basis of shape (the *shape bias*[30]). Importantly, the shape bias only appears around 18–24 months. The psychologist Charles Kemp and his collaborators have argued that the shape bias is acquired through hierarchical Bayesian learning; by exposure to multiple word-referent mappings, children learn that objects with common shapes tend to share a common name, and thus learn to rely differentially on this feature in the future.[31] Consistent with the hierarchical learning account, 17-month-olds can be trained to exhibit the shape bias through exposure to artificial categories organized by shape.[32]

More direct evidence for hierarchical learning comes from a study by Kathryn Dewar and Fei Xu, whose experimental design is schematized in Figure 3.9. Dewar and Xu showed nine-month-olds a series of objects drawn from different boxes. The shapes of the objects were consistent within a box, but their colors varied. According to the hierarchical learning account, this should lead to a "second-order" generalization that objects from a new box should have a consistent shape, even in the absence of any exposure to objects from the box. To test this hypothesis, infants in the experimental condition were shown two objects from a new box that either matched in shape (expected outcome) or mismatched (unexpected outcome). Consistent with the hypothesis that their second-order generalization was violated, infants looked longer at the unexpected outcome, despite the fact that they had already seen it sampled from an earlier box and hence might expect it to be less surprising. Importantly, infants in the control condition looked equally long at the same unexpected outcome when it was drawn from one of the previously sampled boxes, indicating that the looking-time result is not

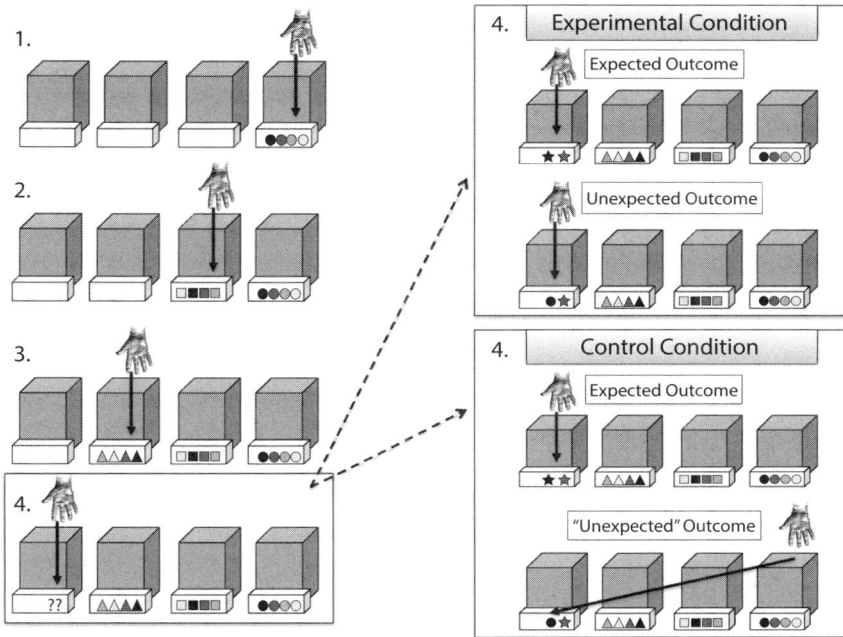

FIGURE 3.9. Experimental design. From Dewar and Xu (2010). Copyright © 2010 by SAGE Publications. Reprinted by permission.

simply an effect of "first-order" surprise (seeing an old object sampled from a new box).

Structure learning

A particularly important form of hierarchical learning comes into play when people have uncertainty not only about the values of particular parameters in the environment, but also about the very structure of the environment. This is the problem of structure learning: What's out there?

The problem is nicely illustrated by an episode in the life of Marco Polo, the Venetian merchant who traveled across Asia in the late 13th century. During his travels in Java, he encountered a rhinoceros. Lacking the concept of a rhinoceros (it had yet to be widely reported in Europe), Marco Polo was faced with a dilemma: Is this a strange exemplar of a familiar category (unicorns) or a fundamentally new category? Although he ultimately decided it was a unicorn, he acknowledged that "they are not of that description of animals which suffer themselves to be taken by maidens, as our people suppose, but are quite of a contrary nature."[33]

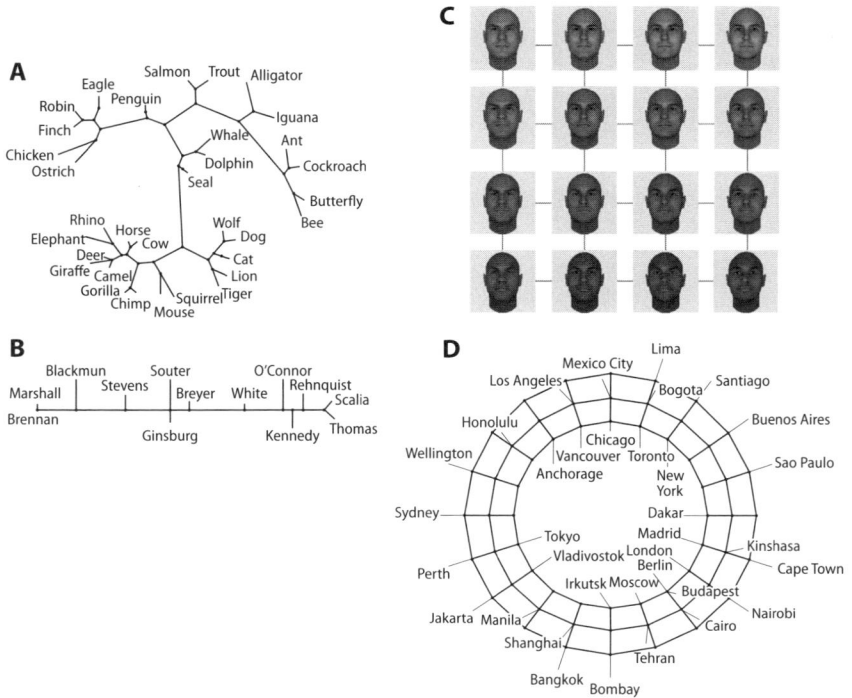

FIGURE 3.10. Examples of structural forms inferred by Kemp and Tenenbaum's Bayesian inference algorithm given different inputs: (A) biological features; (B) Supreme Court votes; (C) face similarities; (D) city distances. Nodes correspond to objects in a domain, and edges correspond to probabilistic dependencies between the nodes. Reproduced from Kemp and Tenenbaum (2008).

Marco Polo's problem is our problem: we are not equipped at birth with concepts like rhinoceros, telephone, airplane, etc., so we have to discover them *de novo*. This raises the question of what kinds of inductive biases we have for structure. The space of possible structures is infinite, but evidence suggests that we have a preference for simpler structures.[34] Whenever possible, we reduce complex perceptual stimuli into a smaller number of "clusters" (e.g., groups of dots) or "features" (e.g., points that join together into lines, and lines joined into shapes) and we even apply this reductionist approach to the discovery of social groups.[35] For example, the degree of social influence that another person exerts on your choices depends not only on your similarity with that person (e.g., how often you agree on past choices) but also on the degree to which you believe that the person belongs to the same social group as you. This belief can be influenced by observing that this person's choices cluster together with the choices of others besides you, such that it is plausible that these individuals form a group. The group identity is "latent" in the sense that there are no external labels that define the group; it may be purely a construction of your mind.

The language of structural forms is much richer than just clusters and features. Charles Kemp and Josh Tenenbaum have developed a model of structure learning based on "graph grammars" that can produce a dazzling variety of structural forms, such as grids, rings, chains, and trees.[36] Given some data, a Bayesian inference algorithm can be used to estimate which graph is the most suitable structure (Figure 3.10). For example, when given Supreme Court votes as input, their algorithm infers that the justices lie along a line (liberal to conservative). Biological features are mapped to a taxonomic tree, faces are mapped to a grid, distances on a circular scheme. Kemp and Tenenbaum argued that graph grammars offer a plausible inductive bias for structure learning, since they are expressive enough to capture much of the diversity of human knowledge, but are still reasonably constrained relative to the space of all possible structures. That being said, graph grammars clearly do not exhaust the richness of human inductive biases for structural knowledge, and more recent work has suggested that ideas from programming languages furnish more sophisticated inductive biases that would (for example) enable the discovery of recursive concepts (e.g., plants, stairs, and numbers that lack a fixed set of features but can be specified recursively).[37]

Summary

Evidence suggests that people employ inductive biases organized around causality, compositionality, objects, and agents, which provide the building blocks for rich, domain-specific intuitive theories. Some aspects of these inductive biases may be acquired through a process of hierarchical learning, which formalizes the notion that people not only learn about the world—they also learn how to learn by means of abstract, higher-order generalizations.

4

Learning from others

Tradition is not the cult of ashes, it is the transmission of fire.

—GUSTAV MAHLER

In his book *The Secret of Our Success*, the anthropologist Joe Henrich tells the tales of two shipwrecked expeditions that met very different fates—one perishing, the other surviving—despite comparable training, technology, and supplies.[1] What made the difference? Henrich argues that the survivors did not have superior intellect, only superior social skills. Unlike the expedition that perished, the survivors befriended the local Inuit and learned their practices. Henrich's point is that big brains on their own are not very useful; our key evolutionary advantage is the ability to learn from one another. Henrich summarizes his thesis as follows:

> The secret of our species' success resides not in the power of our individual minds, but in the *collective brains* of our communities. Our collective brains arise from the synthesis of our culture and social natures— from the fact that we readily learn from others (are *cultural*) and can, with the right norms, live in large and widely interconnected groups (are *social*).[2]

Henrich's thesis is supported by a wide range of cross-cultural and cross-species studies.

Even granting that social learning is the secret of our success, psychologists have documented many apparently stupid social learning behaviors. In this chapter, I take a closer look at some of these behaviors, and make the argument that they are more intelligent than they appear.

Conformity and information cascades

In a seminal experiment, the psychologist Solomon Asch showed people a set of lines and asked them which of the lines matched a probe line.[3] When people were alone, they made virtually no errors. Asch then examined what would happen when people had to make the choice in the presence of other people making the same choice. Unbeknownst to his experimental participants, the other people in the experiment were confederates who occasionally and unanimously chose the wrong match. Asch observed that people then made systematic errors, siding with the majority judgment despite the "obvious" incorrectness of this judgment.

At first blush, this seems like social learning run amok: we ignore the correct data from our senses in favor of incorrect data from other people. But as we've already seen, biases are not necessarily irrational—they are in fact essential for efficient learning. In the real world, the judgments of other people typically provide useful information, so using that information should lead to higher accuracy on average, even if in some cases (the Asch experiment being an extreme) this leads to lower accuracy.

The rationality of conformity has been studied by economists using the idea of an *information cascade*.[4] Individuals sequentially receive both private information (e.g., visual information about line length, as in the Asch experiment) and public information (e.g., line length judgments from individuals earlier in the sequence). The key result is that integrating these two sources of information rationally using Bayes' rule and reporting the optimal judgment can cause subsequent individuals to ignore their private information in favor of the public information. When this happens, the sequence is said to be in a *cascade*. Even when the private information is fairly reliable, the sequence can produce incorrect judgments (i.e., an incorrect cascade); once an individual ignores their private information, their reports do not convey any new information to subsequent individuals. In general, the probability of not being in a cascade (either correct or incorrect) falls exponentially as the number of individuals in the sequence grows. This means that large information networks will with high probability lead to Asch-like neglect of private information (though fortunately the probability of being in a correct cascade typically also increases with the number of individuals). Conformity is thus not a niche phenomenon, but rather a fundamental consequence of how information propagates when individuals cannot access each other's private information.

The theory of information cascades has extremely broad implications for conformist behavior in many domains.[5] It offers an explanation of why some erroneous medical practices (e.g., bleeding) can persist for centuries, why market bubbles form, and why small protests can transform into revolutions. It also suggests a rationale for certain widespread marketing practices, such as setting low initial prices on products. If purchasers form an information cascade, then even a slight bias early in the cascade to purchase a particular product can mushroom into a large bias later in the cascade. Similarly, initial public offerings of equity are typically underpriced, possibly to encourage the formation of an information cascade.[6]

There is evidence that some organizations take into account information cascades and try to mitigate their potentially negative consequences.[7] Many military courts have officers vote on cases in order of increasing rank, in order to reduce the influence of high-ranked officers on low-ranked officers. In some countries, polling immediately prior to elections is outlawed in order to prevent the poll results from influencing the judgments of individual voters. The same reasoning applies to the design of simultaneous voting systems, which prevent individuals from knowing each other's votes.

Overimitation

Imitation is a powerful strategy for social learning: if you observe someone successfully solve a problem, simply copy (or try to copy) the solution. Of course, this can sometimes lead you astray if the person you're observing is not very competent, or if you subsequently discover new information that indicates a better solution. In these cases, a wiser strategy is to emulate rather than to imitate: observe the outcomes of a person's actions and then try to reproduce the outcome, which might entail applying actions that are different than the ones you observed.

Much to the embarrassment of humankind, it turns out that wild chimpanzees are better emulators than three- and four-year-old children.[8] To demonstrate this, experimenters presented chimpanzees and children with a transparent puzzle box (Figure 4.1), demonstrating a series of actions that culminated in the removal of a prize from the box. Critically, some of the actions were causally unnecessary, whereas others were causally necessary. When allowed to freely play with the box, chimpanzees ignored the unnecessary actions and only performed the necessary ones. In other words, they emulated the experimenter. Children, in contrast, performed both the unnecessary and necessary actions—they *overimitated* the experimenter. It is not the case that chimpanzees simply didn't know how to imitate the experimenter, because when the puzzle box was opaque instead of transparent, the chimpanzees (like the children) performed the unnecessary actions. In this case, neither the chimpanzees nor the children had access to information about a more efficient solution.

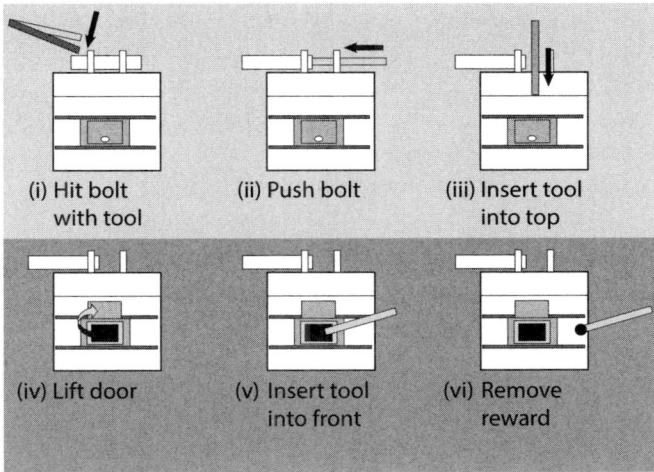

FIGURE 4.1. Apparatus used in the experiments of Horner and Whiten (2005). The top row shows causally unnecessary actions; the bottom row shows causally necessary actions. Figure adapted from Lyons et al. (2006).

In fact, overimitation is not limited to children—adults overimitate too, including in naturalistic settings without any intentional demonstrations.[9] Adults may even overimitate *more* than children in some circumstances.[10] So what is going on here? Why are humans engaging in this manifestly stupid observational learning behavior?

Of course, humans are perfectly capable of emulation. We do it all the time. The question is why we overimitate in the particular circumstances that have been studied experimentally (e.g., with unfamiliar puzzle boxes). The answer is that necessity from the experimenter's point of view is not the same as necessity from the observer's point of view. Once we recognize that necessity is ambiguous and must be inferred on the basis of available information, a rational account of overimitation comes into view. Consistent with this account, experiments have shown that overimitation is actually quite sophisticated. Children are sensitive to the contexts in which unnecessary actions are performed, including the mental states of the actor, and adjust their overimitation based on inferences about the necessity of actions.

Before discussing some of these experimental results, it will be helpful to articulate a theoretical framework that explains them. The psychologist Noah Goodman and his collaborators developed a framework in which the causal structure of the problem is known to the demonstrator but not to the observer.[11] The demonstrator chooses a sequence of actions that maximizes their net utility (obtained reward minus the cost of action). Intuitively, the demonstrator will choose the shortest sequence of actions that solves the task. The observer can use information provided by the demonstration, combined with the assumption that the observer is both knowledgeable and efficient, to infer (using Bayes' rule) the underlying causal structure.

For example, consider the case where action sequence AB yields a reward. Without the knowledgeable/efficient observer assumption, the inference problem is fundamentally ambiguous: one or both actions might be necessary to obtain a reward (this is known as *causal confounding*). But if the observer knows that the demonstrator is both knowledgeable and efficient, then it follows from Bayesian inference that both actions are necessary; otherwise there would be no reason for the demonstrator to pay the cost of action. This prediction was verified experimentally: adults regarded AB as necessary to obtain a reward when they knew that the demonstrator was knowledgeable, but not when the demonstrator was ignorant. Similar results have been reported in three- to five-year-old children.[12]

Goodman and colleagues suggested that the difference between chimpanzees and humans is not the nature of their reasoning but the nature of the mental models over which reasoning operates. In particular, chimpanzees might be able to reason probabilistically about causal structure, but they might not be able to use social information to guide their reasoning. This view dovetails with the important observation that overimitation does not seem to emerge until around three or four years of age, roughly consistent with the developmental trajectory of mental state reasoning.[13] For example, three-year-olds have trouble understanding the false beliefs of other agents; they appear to believe that beliefs must always be consistent with the world.[14] More broadly, three-year-olds have trouble understanding the nature of mental representations, including their own, demonstrated by the fact that they struggle to remember and report past mental states.[15]

Several other lines of evidence support the view that overimitation reflects rational reasoning over mental states.[16] First, children are less likely to overimitate when actions result in a costly outcome (e.g., destruction of an object known to be valuable to the demonstrator), or when the demonstrator emphasizes the instrumental nature of the actions by announcing their goal.[17] Second, children will identify some unnecessary actions as goals, provided the actions are perceived as intentional.[18] Finally, children will overimitate less if the unnecessary actions are performed in a context different from the one in which the initial demonstration took place, consistent with the hypothesis that children consider the conditions for action initiation.[19]

Advice taking

While overimitation reflects the high level of trust we typically have in the behavior of other people, studies of advice taking find the opposite pattern: underutilization of advice (also known as an *egocentric bias*).[20] It is important to note that underutilization is defined with respect to the objectively correct answer, which is of course unknown to participants in these studies. Thus, we again must ask the question, how *should* people rationally update their beliefs when they receive advice?

Much like in multi-sensory cue combination (see Chapter 2), Bayes' rule tells us that we should weigh advice in proportion to an advisor's reliability relative to our own reliability.[21] This is consistent with experimental studies, which have shown that utilization of advice increases with the advisor's reliability and decreases with the learner's reliability.[22] Quantitatively, people appear to rely on advice more than predicted by Bayesian inference. In other words, people place more weight on advice than is mandated by the relative reliability of the advisor (compared to their own reliability). Thus, although people underutilize advice compared to the objectively correct answer, they actually *overutilize* advice compared to rational belief updating.

Another striking observation about advice taking is that people tend to discount advice more when it is far from their own initial opinion.[23] This seems somewhat paradoxical, since you might imagine that rational learners should update to a greater degree when they receive discrepant advice. However, it makes sense when you consider that reliability of the advisor must be inferred from the available data. As the advice becomes more discrepant, you might begin to have the suspicion that the advisor doesn't know what they're talking about.[24] Critically, this depends on how confident you are in your own knowledge, since naturally you might not know what you're talking about either. This line of reasoning explains why discounting of highly discrepant advice is stronger when the learner is knowledgeable,[25] and weaker when the advisor is knowledgeable.[26]

The Bayesian account of advice taking sheds light on another surprising finding: sometimes learners will paradoxically change their opinions in the direction *opposite* of the advisor, particularly when they infer (or are told) that the advisor has low reliability.[27] According to a Bayesian analysis developed by Daniel Hawthorne-Madell and Noah Goodman, this occurs because low-reliability advice can actually provide positive evidence for the alternative ("that guy is always wrong; listen to his advice and then do the opposite").[28]

Paradoxical effects of advice are widespread. Many have in common an explanation in terms of inference about hidden causes.[29] For example, the anti-drug campaign D.A.R.E. (Drug Abuse Resistance Education) encouraged American children to "Just say no" to peer pressure about taking drugs. However, some evidence indicates that alcohol consumption and cigarette smoking may have actually *increased* among its target audience.[30] One interpretation of this outcome is that the "Just say no" slogan implicitly provided evidence that drug-taking is popular among peers and thus has high value.

Summary

Human social learning abilities, starting at a young age, are remarkably sophisticated, incorporating inferences about mental states, causal structure, and reliability of social information. Yet it is precisely this sophistication that gives rise to apparently maladaptive behaviors like conformist errors, overimitation, and underutilization of advice. Understanding the deeper logic of social learning reveals that these maladaptive behaviors are the consequences of generally adaptive biases.

5

Good questions

Judge a man by his questions, not by his answers.
—VOLTAIRE

The previous chapters have treated learners as passive recipients of information, but in reality people have a great deal of agency over the information they receive. This raises the normative question of what information people should seek—what questions should they ask about the world? We will begin with a well-known bias to ask questions that will confirm hypotheses rather than falsify them. We will then consider the deeper structure underlying this behavior within the framework of optimal information acquisition (otherwise known as *active learning*). This will lead us to the conclusion that confirmation bias can in fact be rational under some conditions, including those in which it has been experimentally demonstrated, and that people adopt a disconfirmation strategy under other conditions.

The positive test strategy

The term *confirmation bias* is often used to embrace a number of distinct phenomena.[1] One important distinction is between how people *select* information and how they *interpret* information. In the next chapter, I will discuss the tendency to ignore information that appears to discredit one's hypotheses. In this chapter, I will focus on the tendency to select information that confirms one's hypotheses, sometimes referred to as the *positive test strategy*, where a "test" is a question that one asks about the world in order to assess the validity of a hypothesis, and tests are "positive" when they tend to reveal confirmatory information.[2]

The classic experimental work on the positive test strategy was carried out by the psychologist Peter Wason using a card-selection task.[3] In this task,

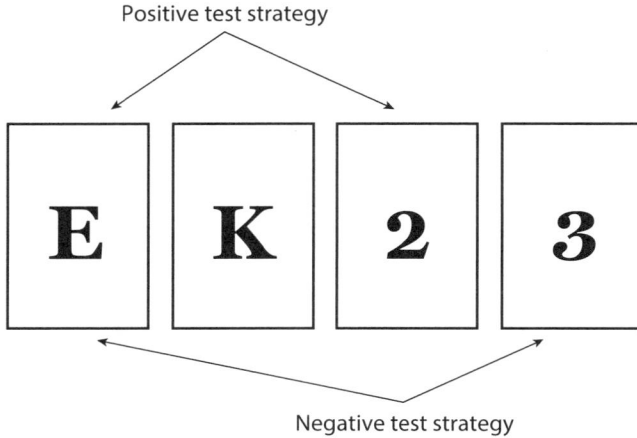

FIGURE 5.1. Cards used in the Wason selection task.

people are shown four cards (Figure 5.1), each with a number on one side and a letter on the other side, and are asked which cards they must turn over to determine the validity of the rule "If there is a vowel on one side, then there is an even number on the other side." Logically, the correct answer is to turn over the vowel card and the odd card. Finding an odd number behind the vowel or a vowel behind the odd number would reveal the hypothesis to be false. In other words, the logically correct answer follows a *falsificationist* (negative test) strategy, as advocated most famously by the philosopher Karl Popper.[4] By contrast, turning over the consonant card or the even card cannot falsify the hypothesis; one can find a pattern *consistent* with the rule, but that still leaves open the possibility that the rule is false in general.

Most people do not follow the negative test strategy. The vast majority either only turn over the vowel card, or turn over the vowel and even cards. In other words, people seem to be following a positive test strategy, turning over cards that confirm the rule. This has been taken as evidence that people are fundamentally irrational (from a logical perspective) when testing hypotheses.

Optimal information acquisition

Let's take another look at the rationality of the positive test strategy. What makes a question good? The standard logical analysis of the selection task assumes that a question is good if it can falsify the hypothesis in question, but this is not the only way to think about the problem. We can alternatively ask what questions are *maximally informative*. It turns out that informative questions are not the same as falsifying questions.

Intuitively, a datum is informative to the extent that it reduces your uncertainty about a variable that you cannot directly observe. If I hear a loud crash in the room where my kids are playing, I'll usually ask my four-year-old what happened, rather than my one-year-old, since her answer will surely reduce my uncertainty to a greater degree. To measure uncertainty reduction quantitatively, we have to formalize uncertainty (Box 5.1). Suppose we observe random samples of a variable drawn from a probability distribution. The mathematician Claude Shannon formalized uncertainty as how surprised we are by these samples on average, where surprise is measured by the negative log probability of the samples, sometimes referred to by the technical term *surprisal*.[5] Using this definition, average surprise corresponds to the *entropy* of the distribution. When the variable is perfectly predictable (all samples are the same), entropy is 0. The entropy is highest when the samples are maximally unpredictable. The entropy is equivalent to the average number of yes/no questions one would need to ask in order to most efficiently ascertain the identity of a single sample.[6] For example, if the variable is binary (a coin flip), then you would need to ask one question on average (entropy = 1) for a fair coin: Did it land heads or tails? If the coin is bent so that it always lands tails, then you wouldn't need to ask any questions (entropy = 0).

Box 5.1. Some concepts from information theory

Let X denote a random variable with probability distribution $P(X)$. The *entropy* of X is given by:

$$H(X) = - \sum_x P(x) \log P(x), \qquad (5.1)$$

where I have used lower case to denote particular values of the random variable. The mutual information between random variables X and Y is given by:

$$I(X; Y) = H(X) - H(X|Y), \qquad (5.2)$$

where $H(X|Y)$ is the average surprise after observing Y (the *conditional entropy*):

$$H(X|Y) = - \sum_y P(y) \sum_x P(x|y) \log P(x|y). \qquad (5.3)$$

Notice that the conditional entropy also includes an average over y, since we want to know how informative samples of y are about x on average.

Now that we have defined uncertainty quantitatively, we can define uncertainty reduction as the difference in entropy about a hypothesis before and after some data are observed. This difference is known as the *mutual information* between the hypothesis and data variables. Mutual information can also be interpreted as "Bayesian surprise" —how much the posterior over hypotheses diverges on average from the prior over hypotheses.[7]

We now can see that the optimal information-acquisition strategy for uncertainty reduction is to select the test with highest mutual information.[8] This is sometimes known as the *expected information gain* strategy. With this mathematical formalism in hand, let us return to the Wason selection task and re-examine the rationality of human performance.

When is the positive test strategy optimal?

Applying the expected information gain strategy to the Wason selection task, the psychologists Mike Oaksford and Nick Chater made the critical observation that following the positive test strategy (selecting the vowel and even cards) is optimal when vowels and even numbers are relatively rare in the environment.[9] They referred to this as the *rarity assumption*, which they justified by arguing that most properties for which we have words refer to events and objects that are rare in the environment.[10] I will discuss the evidence for this assumption further in the next chapter.

Oaksford and Chater demonstrated the versatility of their theory by applying it to a number of different variations of the selection task. For example, the theory correctly predicts that the preference for the negative test (odd card) grows with the probability of observing a vowel or even number (i.e., when the rarity assumption is violated).[11] The theory also explains the effect of exposing people to inconsistencies between their card selections and their independent judgments of whether a card can falsify the rule.[12] Not surprisingly, this exposure encourages them to shift (partially) towards a negative test strategy (choosing the odd card). More surprisingly, it also increases the probability that they will choose the even card. In other words, people exhibit a peculiar mix of negative and positive test strategies after exposure to their inconsistencies. Oaksford and Chater point out that this makes sense under the information-theoretic analysis, because choosing the even card will typically be more informative than choosing the odd card, provided the rarity assumption is satisfied. Thus, the theory predicts that one cannot reasonably adopt a negative test strategy without committing oneself also to a positive test strategy.

Another condition for the rationality of the positive test strategy is when the hypotheses are deterministically related to data.[13] I'll discuss the psychological plausibility of the determinism assumption in the next chapter, but for now it suffices to point out that this assumption, when combined with the expected information-gain strategy, can accurately predict how people

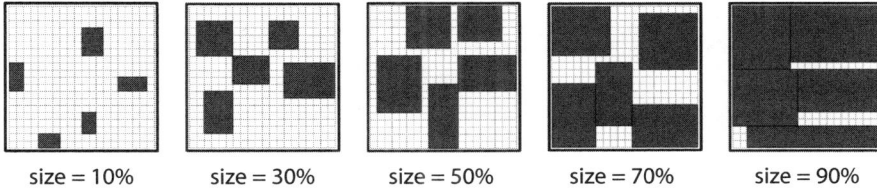

| size = 10% | size = 30% | size = 50% | size = 70% | size = 90% |

FIGURE 5.2. Grey rectangles show ships of various sizes. The task of subjects in this experiment is to place the ships in the correct location. Reproduced from Hendrickson et al. (2016).

test hypotheses in rule-learning tasks (where the assumption of deterministic rules seems natural). The determinism assumption is in fact closely related to the rarity assumption, in the sense that both imply that a given hypothesis is "small"—it can only generate a relatively small number of data patterns, such that observing any of these patterns is highly informative about the underlying hypothesis. Under these conditions, positive evidence for a particular hypothesis is inconsistent with many other hypotheses, and thus should be rationally preferred to negative evidence, which may be consistent with multiple hypotheses.[14] For example, knowing that a creature has black and white stripes is strong positive evidence that it is a zebra, because very few alternative hypotheses could have generated the same data, whereas observing that a creature has two legs is weak negative evidence, because it is consistent with many different hypotheses, even though it rules out the zebra hypothesis.

Some experiments have assessed the usage of the positive test strategy while manipulating hypothesis size. In one experiment, people played a version of the game *Battleship*, in which they were tasked with locating ships on a grid (Figure 5.2).[15] They could request positive evidence (grid squares where ships are located) or negative evidence (grid squares where ships are not located). Requests for positive evidence declined as the size of the hypotheses (literally the size of the ships) increased, consistent with the optimal information-acquisition strategy.

Preferences for and against information

So far, we have been concerned with how people select questions to gain information about hypotheses. But there are further challenges to a theory of optimal information acquisition. Sometimes people want information that is apparently useless, and other times people avoid information that is potentially useful. For example, people will sometimes opt for medical tests even when the outcome of the test will not influence their subsequent medical decisions,[16] or will wait until they obtain information before making a decision, even if this information will not change their decision.[17] People are also willing to pay considerable amounts of money for useless information about lottery outcomes.[18] Yet people will avoid information about the gender of their child before birth,[19] about whether another person got a better deal on an expensive purchase,[20] or about undesirable medical test outcomes,[21] among many other examples.[22] How can we reconcile these seemingly opposite preferences for information?

One prominent account, developed by the economists Andrew Caplin and John Leahy, posits that people assign utilities not only to external states of the world but also to internal states of mind. In particular, uncertainty can elicit emotions like anxiety and anticipation that are captured by a utility function defined over beliefs.[23] Thus, people will delay kissing (hypothetically) their favorite movie star in order to prolong anticipatory pleasure, whereas they will hasten the receipt of painful shocks in order to abbreviate anticipatory displeasure.[24] While this approach can capture some forms of information preferences, it still begs the question *why* the utility function is structured in this way.

The computer scientist Emma Pierson and psychologist Noah Goodman have offered a more fundamental answer to this question, grounded in the notion that planning is cognitively costly, so preferences for information should be sensitive to the expected costs and benefits of planning.[25] Pierson and Goodman analyzed the following stylized scenario: suppose you are locked in a room for an hour and you are told that after the hour has elapsed you will win or lose an amount of money D with probability p (or with probability $1 - p$ nothing will happen). You can pay a cost C when you enter the room to find out whether you will win or lose the money, but this information has no effect on the outcome. Should you pay for the information?

A perfect (cognitively unbounded) Bayesian agent should never pay for the information, because it is instrumentally useless. However, a cognitively bounded agent could use its time in the room to plan (e.g., figure out what to do with the money). But since planning is costly, it only makes sense to plan if the outcome probability p is high, and more generally, the value of information increases monotonically with p if the information cost is 0. However, if the information cost is greater than 0, then the value of information is nonmonotonic: it will increase with p up to a point, and then start to decrease. The reason is that when p is close to 1, then you know (with or without the information) that the event will occur, and hence you'll have to plan either way, so there's no point paying money for the information. Consistent with this analysis, Pierson and Goodman found that people's willingness to pay for information increased monotonically with p when information was costless, but was an inverted-U as a function of p (first increasing and then decreasing) when information was costly.

Summary

When hypotheses are sparse or deterministically related to observations, the confirmation bias is in fact not a bias at all—it is the optimal information-acquisition strategy. In the next chapter, we will see that the same assumptions about the world produce resistance to disconfirmatory evidence. Thus, preference for confirmation and resistance to disconfirmation may be two sides of the same coin, reflecting the structure of how people represent the world.

In addition to addressing the confirmation bias, this chapter discussed preferences for and against information, showing how cognitively bounded rational agents will exhibit information preferences that have previously been deemed irrational. In Chapter 10, I will discuss cognitive resource constraints more broadly, exploring how they allow us to make sense of numerous puzzling behaviors that are difficult to reconcile with a notion of cognitively unbounded rationality.

6

How to never be wrong

No theory ever agrees with all the facts in its domain, yet it is not always the theory that is to blame. Facts are constituted by older ideologies, and a clash between facts and theories may be proof of progress.

—PAUL FEYERABEND

Since the discovery of Uranus in 1781, astronomers were troubled by certain irregularities in its orbit, which appeared to contradict the prevailing Newtonian theory of gravitation.[1] Then, in 1845, Le Verrier and Adams independently completed calculations showing that these irregularities could be entirely explained by the gravity of a previously unobserved planetary body. This hypothesis was confirmed a year later through telescopic observation, and thus an eighth planet (Neptune) was added to the solar system. Le Verrier and Adams succeeded on two fronts: they discovered a new planet, and they rescued the Newtonian theory from disconfirmation.

Neptune is a classic example of what philosophers of science call an ad hoc auxiliary hypothesis.[2] All scientific theories make use of auxiliary assumptions that allow them to interpret experimental data. For example, an astronomer makes use of optical assumptions to interpret telescope data, but one would not say that these assumptions are a core part of an astronomical theory; they can be replaced by other assumptions as the need arises (e.g., when using a different measurement device), without threatening the integrity of the theory. An auxiliary assumption becomes an ad hoc hypothesis when it entails unconfirmed claims that are specifically designed to accommodate disconfirmatory evidence.

Ad hoc auxiliary hypotheses have long worried philosophers of science, because they suggest a slippery slope towards unfalsifiability.[3] If any theory can be rescued in the face of disconfirmation by changing auxiliary assumptions, how can we tell good theories from bad theories? While Le Verrier and Adams were celebrated for their discovery, many other scientists were less fortunate. For example, in the late 19th century, Michelson and Morley reported experiments apparently contradicting the prevailing theory that electromagnetic radiation is propagated through a space-pervading medium (ether). FitzGerald and Lorentz attempted to rescue this theory by hypothesizing electrical effects of ether that were of exactly the right magnitude to produce the Michelson and Morley results. Ultimately, the ether theory was abandoned, and Karl Popper derided the FitzGerald-Lorentz explanation as "unsatisfactory" because it "merely served to restore agreement between theory and experiment."[4]

Ironically, Le Verrier himself was misled by an ad hoc auxiliary hypothesis. The same methodology that had served him so well in the discovery of Neptune failed catastrophically in his "discovery" of Vulcan, a hypothetical planet postulated to explain excess precession in Mercury's orbit. Le Verrier died convinced that Vulcan existed, and many astronomers subsequently reported sightings of the planet, but the hypothesis was eventually discredited by Einstein's theory of general relativity, which accounted precisely for the excess precession without recourse to an additional planet.

The basic problem posed by these examples is how to assign credit or blame to central hypotheses vs. auxiliary hypotheses. An influential view, known as the Duhem-Quine thesis (reviewed in the next section), asserts that this credit-assignment problem is insoluble—central and auxiliary hypotheses must face observational data "as a corporate body."[5] This thesis implies that theories will be resistant to disconfirmation as long as they have recourse to ad hoc auxiliary hypotheses.

Psychologists recognize such resistance as a ubiquitous cognitive phenomenon, commonly viewed as one among many flaws in human reasoning.[6] However, as the Neptune example attests, such hypotheses can also be instruments for discovery. The purpose of this chapter is to discuss how a Bayesian framework for induction deals with ad hoc auxiliary hypotheses,[7] and then to leverage this framework to understand a range of phenomena in human cognition. According to the Bayesian framework, resistance to disconfirmation can arise from rational belief updating mechanisms, provided that an individual's "intuitive theory" satisfies certain properties: a strong prior belief in the central hypothesis, coupled with an inductive bias to posit auxiliary hypotheses that place high probability on observed anomalies. The question then becomes whether human intuitive theories satisfy these properties, and several lines of evidence suggest the answer is yes.[8] In this light, humans are surprisingly rational. Human beliefs are guided by strong inductive biases about the world. These biases enable the development of robust intuitive theories, but can sometimes lead to preposterous beliefs.

Underdetermination of theories: The Duhem-Quine thesis

Theories (both scientific and intuitive) are webs of interconnected hypotheses about the world. Thus, one often cannot confirm or disconfirm one hypothesis without affecting the validity of the other hypotheses. How, then, can we establish the validity of an individual hypothesis? The philosopher Pierre Duhem brought this issue to the foreground in his famous treatment of theoretical physics:

> The physicist can never subject an isolated hypothesis to experimental test, but only a whole group of hypotheses; when the experiment is in disagreement with his predictions, what he learns is that at least one of the hypotheses constituting this group is unacceptable and ought to be modified; but the experiment does not designate which one should be changed.[9]

While Duhem restricted his attention to theoretical physics, W.V.O. Quine took the same point to its logical extreme, asserting that *all* beliefs about the world are underdetermined by observational data:

> The totality of our so-called knowledge or beliefs, from the most casual matters of geography and history to the profoundest laws of atomic physics or even of pure mathematics and logic, is a man-made fabric which impinges on experience only along the edges. Or, to change the figure, total science is like a field of force whose boundary conditions are experience. A conflict with experience at the periphery occasions readjustments in the interior of the field. But the total field is so underdetermined by its boundary conditions, experience, that there is much latitude of choice as to what statements to reevaluate in the light of any single contrary experience. No particular experiences are linked with any particular statements in the interior of the field, except indirectly through considerations of equilibrium affecting the field as a whole.[10]

In other words, one cannot unequivocally identify particular beliefs to revise in light of surprising observations. Quine's conclusion was stark: "The unit of empirical significance is the whole of science" (p. 42).

Some philosophers have taken undetermination to invite a radical critique of theory testing. If evidence cannot adjudicate between theories, then non-empirical forces, emanating from the social and cultural environment of scientists, must drive theory change. For example, the "research programmes" of Imre Lakatos[11] and the "paradigms" of Thomas Kuhn[12] were conceived as explanations of why scientists often stick to a theory despite disconfirming evidence, sometimes for centuries. Lakatos posited that scientific theories contain a hard core of central theses that are immunized from refutation by a "protective belt" of auxiliary hypotheses. According to this view, science does not progress by falsification of individual theories, but rather by developing a *sequence* of theories that progressively add novel predictions, some of which are corroborated by empirical data.

While the radical consequences of underdetermination have been disputed,[13] the problem of credit assignment remains a fundamental challenge for the scientific enterprise. I now turn to a Bayesian approach to induction that attempts to answer this challenge.

The Bayesian answer to underdetermination

Probability theory offers a coherent approach to credit assignment.[14] Instead of assigning all credit to either central or auxiliary hypotheses, probability theory dictates that credit should be apportioned in a graded manner according to the "responsibility" each hypothesis takes for the data (Box 6.1). This formulation is the crux of the Bayesian answer to underdetermination.[15] A Bayesian scientist does not wholly credit either the central or auxiliary hypotheses, but rather distributes the credit according to the marginal posterior probabilities.

Box 6.1. Bayesian confirmation theory

Let h denote the central hypothesis (e.g., Newton's theory of gravitation), a denote the auxiliary hypothesis (e.g., existence of Neptune), and d denote the data (e.g., irregularities in the orbit of Uranus). After observing d, the prior probability of the conjunct ha, $P(ha)$, is updated to the posterior distribution $P(ha|d)$ according to Bayes' rule:

$$P(ha|d) = \frac{P(d|ha)P(ha)}{P(d|ha)P(ha) + P(d|\neg(ha))P(\neg(ha))}, \tag{6.1}$$

where $P(d|ha)$ is the likelihood of the data under ha, and $\neg(ha)$ denotes the negation of ha.

The sum rule of probability allows us to ascertain the updated belief about the central hypothesis, marginalizing over all possible auxiliaries:

$$P(h|d) = P(ha|d) + P(h\neg a|d). \tag{6.2}$$

Likewise, the marginal posterior over the auxiliary is given by:

$$P(a|d) = P(ha|d) + P(\neg ha|d). \tag{6.3}$$

This analysis does not make a principled distinction between central and auxiliary hypotheses: they act conjunctively, and are acted upon in the same way by the probability calculus. What ultimately matters for distinguishing them, as illustrated below, is the relative balance of evidence for the different hypotheses. Central hypotheses will typically be more entrenched due to a stronger evidential foundation, and thus auxiliary hypotheses will tend to be the elements of Quine's "total field" that readjust in the face of disconfirmation.

I will not address here the philosophical controversies that have surrounded the Bayesian analysis of auxiliary hypotheses.[16] My goal is not to establish the normative adequacy of the Bayesian analysis, but rather to explore its implications for cognition—in particular, how it helps us understand resistance to belief updating (Box 6.2).

Box 6.2. An illustrative analysis of belief updating

Following Strevens (2001), I illustrate the dynamics of belief by assuming that the data d have their impact on the posterior probability of the central hypothesis h solely through its falsification of the conjunct ha (i.e., the data indicate that the conjunction of central and auxiliary hypotheses are collectively false),[17] which implies:

$$P(h|d) = P(h|\neg(ha)). \tag{6.4}$$

In other words, the likelihood is 0 for ha and 1 for all other conjuncts. Under this assumption, Strevens obtains the following expression:

$$P(h|d) = \frac{1 - P(a|h)}{1 - P(a|h)P(h)}P(h). \tag{6.5}$$

This expression has several intuitive properties, illustrated in Figure 6.1. As one would expect, the posterior probability of h always decreases following disconfirmatory data d. The

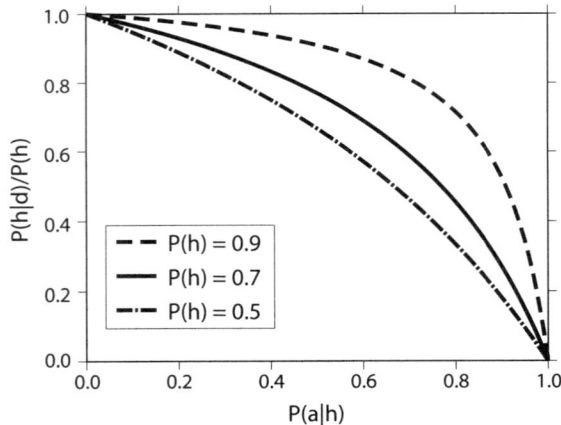

FIGURE 6.1. Ratio of posterior to prior probability of the central hypothesis h as a function of the probability of the auxiliary hypothesis a given h, plotted for three different priors for the central hypothesis. From Strevens (2001). Copyright © 2001 by Oxford University Press. Reprinted by permission.

decrease in the posterior probability is inversely related to $P(h)$ and directly related to $P(a|h)$.[18] Thus, a central hypothesis with high prior probability relative to the auxiliary hypothesis [i.e., high $P(h)/P(a|h)$] will be relatively robust to disconfirmation, pushing blame onto the auxiliary. But if the auxiliary has sufficiently high prior probability, the central hypothesis will be forced to absorb the blame. It is important to see that the robustness to disconfirmation conferred by a strong prior is not a bias due to motivated reasoning (believing things because you want them to be true)—it is a direct consequence of rational inference.[19]

One might wonder how this analysis determines whether an auxiliary hypothesis is ad hoc or not. The answer is that it doesn't: the only distinguishing features of hypotheses are their prior probabilities and their likelihoods. Thus, on this account "ad hoc" is simply a descriptive label that we use to individuate hypotheses that have low prior probability and high likelihoods. By the same token, a "good" versus "bad" ad hoc auxiliary hypothesis is determined entirely by the prior and likelihood.

Robustness of intuitive theories

One strong assumption underlying the analysis in Box 6.2 is worth highlighting, namely that the likelihood of the central hypothesis being true and the auxiliary being false ($h\neg a$), marginalizing over all alternative auxiliaries (a_k), is equal to 1. I will refer to this as the *consistency assumption*, because it states that only auxiliary hypotheses that are highly consistent with the data will have non-zero probability. Ad hoc auxiliary hypotheses by definition satisfy this assumption. But why should these hypotheses be preferred over others? One way to justify this assumption is to stipulate that there is uncertainty about the parameters of the distribution over auxiliary hypotheses. The prior over these parameters can express a preference for redistributing probability mass (i.e., assigning credit) in particular ways once data are observed. In addition to sparsity, the consistency assumption requires *deterministic* hypotheses.[20] In summary, sparsity and determinism jointly facilitate the robustness of theories. I will argue that these properties characterize human intuitive theories.

Sparsity

The sparsity assumption—that only a few auxiliary hypotheses have high probability—has appeared throughout cognitive science in various guises. Klayman and Ha (1987) posited a *minority phenomenon* assumption, according to which the properties that are characteristic of a hypothesis tend to be rare. For example, AIDS is rare in the population but highly correlated with HIV; hence, observing that someone has AIDS is highly informative about whether they have HIV. This assumption has been invoked to justify the "positive test strategy" prevalent in human hypothesis testing.[21] If people seek confirmation for their hypotheses, then failure to observe the confirmatory evidence will provide strong evidence against the hypothesis under the minority phenomenon assumption. As discussed in the last chapter, Oaksford and Chater used the same idea (what they called the *rarity assumption*) to explain the use of the positive test strategy in the Wason card-selection task.[22] Violations of the sparsity assumption, or contextual information that changes perceived sparsity, causes people to shift away from the positive test strategy.[23] Experiments on hypothesis evaluation tell a similar story: the evidential impact of observations is greater when they are rare,[24] consistent with the assumption that hypotheses are sparse.

Beyond hypothesis testing and evaluation, evidence suggests that people tend to generate sparse hypotheses when presented with data. When participants were asked to generate hypothetical number concepts applicable to the range [1,1000], most of these hypotheses were sparse.[25] For example, a common hypothesis was prime numbers, with a sparsity of 0.168 (i.e., 16.8% of numbers in [0,1000] are primes). Overall, 83% of the generated hypotheses had a sparsity level of 0.2 or less.[26]

Sparsity has also figured prominently in theories of perception. The tuning properties of receptive fields in primary visual cortex may be partially accounted for by assuming that they represent a sparse set of image components.[27] Similar sparse coding ideas have been applied to auditory[28] and olfactory[29] cortical representations. Psychologists have likewise posited that humans parse complex objects into a small set of latent components with distinctive visual features.[30]

Is sparsity a reasonable assumption? Navarro and colleagues attempted to answer this question by demonstrating that (under some fairly generic assumptions) sparsity is a consequence of *family resemblance*: hypotheses tend to generate data that are more similar to one another than to data generated by other hypotheses.[31] For example, members of the same natural category tend to have more overlapping features relative to members of other natural categories.[32] Navarro further showed that natural categories are empirically sparse. Thus, the sparsity assumption may be inevitable if hypotheses describe natural categories.

Determinism

The determinism assumption—that hypotheses tend to generate data near-deterministically—is well-supported as a property of intuitive theories. Some of the most compelling evidence comes from studies of children showing that children will posit a latent cause to explain surprising events, rather than attribute the surprising event to inherent stochasticity.[33] For example, the psychologist Laura Schulz and her collaborators presented four-year-olds with a stochastic generative cause and found that the children inferred an inhibitory cause to "explain away" the stochasticity.[34] Children also expect latent agents to be the cause of surprising motion events, even in the absence of direct evidence for an agent.[35] Like children, adults also appear to prefer deterministic hypotheses.[36] The prevalent use of the positive test strategy in information selection (detailed in Chapter 8) has also been justified using the determinism assumption.[37]

The psychologist Honjing Lu and colleagues have proposed a "generic prior" for causal strength that combines the sparsity and determinism principles.[38] A priori, causes are expected to be few in number and potent in their generative or preventative effects. Lu and colleagues showed quantitatively that this prior, when employed in a Bayesian framework for causal induction, provides a good description of human causal inferences.[39] Buchanan and collaborators developed an alternative deterministic causal model based on an edge replacement process, which creates a branching structure of stochastic latent variables.[40] This model can explain violations of conditional independence in human judgments in terms of the correlations induced by the latent variables.

In summary, sparsity and determinism appear to be central properties of intuitive theories. These properties offer support for the particular Bayesian analysis of auxiliary hypotheses elaborated above, according to which robustness of theories derives from the ability to explain away disconfirmatory data by invoking auxiliary hypotheses.

Implications

Having established the plausibility of the Bayesian analysis, we now explore some of its implications for human cognition. The central theme running through all of these examples is that the evidential impact of observations is contingent on the auxiliary hypotheses one holds; changing one's beliefs about auxiliary hypotheses will change the interpretation of observations. Thus, observations that appear to contradict a central hypothesis can be "explained away" by changing auxiliary hypotheses, and this change is licensed by the Bayesian analysis under the specific circumstances detailed above. If, as I have argued, intuitive theories have the right sort of properties to support this "protective belt" of auxiliary hypotheses,[41] then we should expect robustness to disconfirmation across many domains.

Before proceeding, it is important to note that many of the phenomena surveyed below can also be explained by other theoretical frameworks, such as motivated cognition.[42] The purpose of this section is not to develop a watertight case for the Bayesian framework—which would require more specific model specifications for different domains and new experiments to test rival predictions—but rather to show that evidence for robustness to disconfirmation does not by itself indicate irrationality; it is possible to conceive of a perfectly rational agent who exhibits such behavior. Whether humans really are rational in this way is an unresolved empirical question.[43]

The theory-ladenness of observation

Drawing a comparison between the history of science and perceptual psychology, Thomas Kuhn argued that observation reports are not theory-neutral: "What a man sees depends both upon what he looks at and also upon what his previous visual-conceptual experience has taught him to see."[44] For example, subjects who put on goggles with inverting lenses see the world upside down, but after a period of profound disorientation lasting several days, their perception adapts and they see the world right side up.[45] Thus, the very same retinal image produces starkly different percepts depending on the preceding perceptual history.

More important for Kuhn's argument are examples where percepts, or at least their semantic interpretations, are influenced by the observer's conceptual framework:

> Looking at a contour map, the student sees lines on paper, the cartographer a picture of a terrain. Looking at a bubble-chamber photograph, the student sees confused and broken lines, the physicist a record of familiar subnuclear events. Only after a number of such transformations of vision does the student become an inhabitant of the scientist's world, seeing what the scientist sees and responding as the scientist does. The world that the student then enters is not, however, fixed once and for all by the nature of the environment, on the one hand, and of science, on the other. Rather, it is determined jointly by the environment and the particular normal-scientific tradition that the student has been trained to pursue.[46]

This is essentially a restatement of the view, going back to Hermann von Helmholtz, that perception is a form of "unconscious inference" or "problem-solving"[47] and formalized by modern Bayesian theories of perception.[48]

There is one particular form of theory-ladenness that will concern us here, where changes in auxiliary hypotheses alter the interpretation of observations. Disconfirmation can be transformed into confirmation (e.g., the example of Neptune), or vice versa. When Galileo first reported his observations of mountains on the moon, the critical response focused not on the observations per se but on the auxiliary assumptions mediating their validity. Since the telescope was an unfamiliar measurement device, the optical theory underlying its operation was not taken for granted. In fact, it was non-trivial even to verify Galileo's observations, because many of the other telescopes available in 1610 were of insufficient quality to resolve the same lunar details observed by Galileo. Thus, it was possible at that time to dispute the evidential impact of Galileo's observations for astronomical theories.[49]

Although Galileo's observations were ultimately vindicated, there are other historical examples in which observations were ultimately discredited. For example, Rutherford and Pettersson conducted similar experiments in the 1920s on the emission of charged particles under radioactive bombardment. Pettersson's assistants observed flashes on a scintillation screen (evidence for emission), whereas Rutherford's assistants did not. The controversy was subsequently resolved when Rutherford's colleague, James Chadwick, demonstrated that Pettersson's assistants were unreliable: they reported indistinguishable rates of flashes even under experimental conditions where no particles could have been emitted. The strategy of debunking claims by undermining auxiliary hypotheses has been used effectively throughout scientific history, from Benjamin Franklin's challenge of Mesmer's "animal magnetism" to the revelation that observations of neutrinos exceeding the speed of light were due to faulty detectors.[50]

It is tempting to see a similar strategy at work in contemporary political and scientific debate. In response to negative news coverage, the Trump administration promulgated the idea that the mainstream media is publishing "fake news"—i.e., reports that are inaccurate, unreliable, or biased. This strategy is powerful because it does not focus on the veracity of any one report, but instead attempts to undermine faith in the entire "measurement device." A similar strategy was used for many years by creationists to undermine faith in evolutionary biology, by the tobacco industry to undermine faith in scientific studies of smoking's health effects, and by the fossil fuel industry to undermine faith in climate science. By "teaching the controversy," these groups attempt to dismantle the auxiliary hypotheses on which the validity of science relies. For example, the release of stolen e-mails from the Climatic Research Unit at the University of East Anglia suggested an alternative auxiliary—selective reporting or manipulation of data—that could explain away evidence for human-induced climate change. Indeed, a subsequent survey of Americans showed that over half agreed with the statements "Scientists changed their results to make global warming appear worse than it is" and

"Scientists conspired to suppress global warming research they disagreed with."[51]

A well-studied form of theory-ladenness is the phenomenon of *belief polarization*: individuals presented with the same data will sometimes update their beliefs in opposite directions. In a classic experiment,[52] Lord and colleagues asked supporters and opponents of the death penalty to read about two fictional studies—one supporting the effectiveness of the death penalty as a crime deterrent, and one supporting its ineffectiveness. Subjects who supported the death penalty subsequently strengthened their belief in the effectiveness of the death penalty after reading the two studies, whereas subjects who opposed the death penalty subsequently strengthened their belief in its ineffectiveness. A large body of empirical work on belief polarization was interpreted by many social psychologists as evidence of irrational belief updating.[53] However, another possibility is that belief polarization might arise from different auxiliary hypotheses about the data-generating process.[54] For example, Jern et al. (2014) showed how the findings of Lord et al. (1979) could be accounted for within a rational Bayesian framework. If participants assume the existence of research bias (distortion or selective reporting of findings to support a preconceived conclusion), then reading a study about the ineffectiveness of the death penalty may strengthen their belief in research bias, correspondingly increasing their belief in the effectiveness of the death penalty. Similarly, Cook and Lewandowsky (2016) demonstrated that beliefs in bias of scientific reporting can lead to discounting of climate change evidence. One lesson to draw from these examples is that effective persuasion requires more than simply conveying information confirming or disconfirming central hypotheses; it requires alteration of the auxiliary hypotheses that refract information, rendering perception theory-laden.

Optimism

Many individuals exhibit a systematic "optimism bias,"[55] overestimating the likelihood of positive events in the future.[56] This bias affects beliefs about many real-world domains, such as the probability of getting divorced or being in a car accident. One of the puzzles of optimism is how it can be maintained; even if we start with initial optimism,[57] why doesn't reality force our beliefs to eventually calibrate themselves?

A clue to this puzzle comes from evidence that people tend to update their beliefs more in response to positive feedback compared to negative feedback.[58] The economists David Eil and Justin Rao dubbed this the "good news–bad news effect."[59] For example, Eil and Rao asked subjects to judge the rank of their IQ and physical attractiveness and then received feedback (a pairwise comparison with a randomly selected subject in the same experiment). While subjects conformed to Bayesian updating when they received positive feedback (i.e., when their rank was better than the comparand), they systematically discounted the negative feedback. Similar results have been found using a variety of feedback types.[60]

One reason people may discount negative feedback is that they wish to blunt its "sting."[61] Consistent with this account, Eil and Rao found that subjects who believed that their ranks were near the bottom of the distribution were willing to pay to avoid learning their true rank. An alternative account, drawing from our Bayesian analysis of auxiliary hypotheses, is that people are being fully Bayesian, but their internal model is different from the one presupposed by Eil and Rao. Specifically, let h denote the hypothesis that a person is "high rank," and let a denote the auxiliary hypothesis that the feedback is "valid" (i.e., from an unbiased source). It is intuitive that subjects might discount negative feedback by positing invalid evidence sources; for example, if a person judges you to be unattractive, you could discount this feedback by positing that this person is exceptionally harsh (judges everyone to be unattractive) or is having a bad day.

Suppose we have two people who have the same prior on validity, $P(a|h)$, but different priors on their rank, $P(h)$. The Bayesian analysis developed above (see Figure 6.1) predicts that the person who assigns higher prior probability to being high rank will update less in response to negative feedback. Consistent with this prediction, individuals with higher dispositional optimism were more likely to maintain positive expectations after experiencing losses in a gambling task.[62] The Bayesian analysis also predicts that two people with different priors on validity but the same priors on rank will exhibit different patterns of asymmetric updating, with the weaker prior on validity leading to greater discounting of negative feedback. In support of this prediction, Gilovich and colleagues found that people who observed an outcome that appeared to have arisen from a statistical "fluke" were more likely to discount this outcome when it was negative, presumably since the feedback was perceived to be invalid.[63] The same kind of discounting can lead to overconfidence in financial markets, where investors are learning about their abilities; by taking too much credit for their gains and not enough for their losses, they become overconfident.[64]

A related phenomenon arises in studies of cheating and lying.[65] When people obtain a favorable outcome through unscrupulous means, they tend to attribute this success to their personal ability. For example, Chance and colleagues administered an IQ test to participants that included an answer key at the bottom so that they could optionally "check their work."[66] Compared to participants who did not have the answer key, those with the answer key not only scored higher, but also predicted (incorrectly) that they would score higher on a subsequent test. One way to interpret this result is that participants had a strong prior belief in their ability, which led them to discard the auxiliary hypothesis that cheating aided their score, thereby inflating estimates of their own ability.

The Bayesian analysis predicts that changing auxiliary assumptions will systematically alter updating asymmetries in response to good and bad news. This prediction was tested using a two-armed bandit paradigm in which subjects played the role of prospectors choosing between two mines in search of gold.[67] Each mine was associated with a fixed probability of yielding gold or rocks. In the "benevolent" condition, the subjects were told that a tycoon would intervene on a proportion of trials, replacing the contents of the mine with gold. Importantly, subjects were not told when the tycoon was intervening; they therefore had to infer whether a reward was the result of the intervention or reflected the true underlying reward probability. Because the tycoon would never replace the gold with rocks (the negative outcome), observing rocks was strictly more informative about the underlying reward probability. Subjects in this case were expected to show a pessimism bias, learning more from negative outcomes than from positive outcomes. In contrast, they were expected to show an optimism bias (learning more from positive outcomes than from negative outcomes) in an "adversarial" condition, where a "bandit" replaced the contents of the mine with rocks on a proportion of trials. Computational modeling of the choice data revealed an overall optimism bias, perhaps reflecting the dispositional factors discussed above, but also showed that subjects altered their bias across conditions in accordance with the Bayesian analysis, learning more from negative outcomes in the benevolent condition compared to the adversarial condition.

Controllability

In settings where people might have some control over their observations, beliefs about rank or personal ability are closely connected to beliefs about controllability.[68] If a utility-maximizing agent believes that the world is controllable, then it is reasonable to assume that positive outcomes are more likely than negative outcomes, and hence negative outcomes are more likely to be explained away by alternative auxiliary hypotheses. For example, if you believe that you are a good test-taker (i.e., you have some control over test outcomes), then you may attribute poor test performance to the test difficulty rather than revising your beliefs about your own proficiency; this attribution is less plausible if you believe that you are a bad test-taker (i.e., you lack control over test outcomes). Thus, controllability is an important auxiliary hypothesis for interpreting feedback, with high perceived controllability leading to optimistic beliefs.[69] The link between controllability and rank can be accommodated within the Bayesian framework, since we model the conditional distribution of the auxiliary hypothesis (controllability) given the central hypothesis (rank). This link is supported by studies showing that mood induction can bring about changes in beliefs about controllability.[70]

This analysis of controllability might provide insight into the psychopathology of asymmetric updating in response to positive and negative feedback. Individuals with depression do not show an optimism bias (so-called "depressive realism"[71]), and this may arise from symmetric (unbiased) updating.[72] One possibility is that this occurs because individuals with depression believe that the world is relatively uncontrollable—the key idea in the "learned helplessness" theory of depression, which implies that they cannot take credit for positive outcomes any more than they can discount negative outcomes.[73] Another possibility is that individuals with depression have a lower prior on rank, which would also lead to more symmetric updating compared to non-depressed individuals.

When placed in objectively uncontrollable situations, people will nonetheless sometimes perceive that they have control. According to the Bayesian analysis, this can arise when it is possible to discount unexpected outcomes in terms of an auxiliary hypothesis (e.g., fluke events, intrinsic variability, interventions by alternative causes) instead of reducing belief in control. As pointed out by Harris and Osman (2012), illusions of control typically arise in situations where cues indicate that controllability is plausible. For example, cues suggesting that one's opponent is incompetent inflate the illusion of control in a competitive setting, possibly by increasing the probability that the poor performance of the opponent is due to incompetence rather than the random nature of the outcomes.[74] Another study showed that giving subjects an action that was in fact disconnected from the sequence of outcomes nonetheless inflated their perception that the sequence was controllable.[75] More generally, active involvement increases the illusion of control, as measured by the propensity for risk taking: Davis et al. (2000) found that gamblers in real-world casinos placed higher bets on their own dice rolls than on others' dice rolls.[76]

The basic lesson from all of these studies is that beliefs about controllability and rank can insulate an individual from the disconfirming effects of negative feedback. This response to negative feedback is rational under the assumption that alternative auxiliary hypotheses (e.g., statistical flukes) can absorb the blame.

The true self

Beliefs about the self provide a particularly powerful example of resistance to disconfirmation. People make a distinction between a "superficial" self and a "true" self, and these selves are associated with distinct patterns of behavior.[77] In particular, people hold a strong prior belief that the true self is good (the central hypothesis h in our terminology). This proposition is supported by several lines of evidence. First, positive, desirable personality traits are viewed as more essential to the true self than negative, undesirable traits.[78] Second, people feel that they know someone most deeply when given positive information about them.[79] Third, negative changes in traits are perceived as more disruptive to the true self than positive changes.[80]

The key question for our purposes is what happens when one observes bad behavior: Do people revise their belief in the goodness of the actor's true self? The answer is largely no. Bad behavior is attributed to the superficial self, whereas good behavior is attributed to the true self.[81] This tendency is true even of individuals who generally have a negative attitude towards others, such as misanthropes and pessimists.[82] And even if people are told explicitly that an actor's true self is bad, they are still reluctant to see the actor as truly bad.[83] Conversely, observing positive changes in behavior (e.g., becoming an involved father after being a deadbeat) are perceived as indicating "self-discovery."[84]

These findings support the view that belief in the true good self shapes the perception of evidence about other individuals: evidence that disconfirms this belief tends to be discounted. The Bayesian framework suggests that this may occur because people infer alternative auxiliary hypotheses, such as situational factors that sever the link between the true self and observed behavior (e.g., he behaved badly because its mother just died). However, this possibility remains to be studied directly.

Stereotype updating

Stereotypes exert a powerful influence on our thinking about other people, but where do they come from? We are not born with strong beliefs about race, ethnicity, gender, religion, or sexual orientation; these beliefs must be learned from experience. What is remarkable is the degree to which stereotypes, once formed, are stubbornly resistant to updating.[85] As Lippmann (1922) remarked, "There is nothing so obdurate to education or criticism as the stereotype" (p. 57).

One possible explanation is that stereotypes are immunized from disconfirmation by flexible auxiliary hypotheses. This explanation fits well with the observation that individuals whose traits are inconsistent with a stereotype are segregated into "subtypes" without diluting the stereotype.[86] For example, one study found that stereotypes were updated more when inconsistent traits were dispersed across multiple individuals rather than concentrated in a few individuals, consistent with the idea that concentration of inconsistencies licenses the auxiliary hypothesis that the individuals are outliers, and therefore do not reflect upon the group as a whole.[87] An explicit sorting task supported this conclusion: inconsistent individuals tended to be sorted into separate groups.[88]

These findings have been simulated by a recurrent connectionist model of stereotype judgment.[89] The key mechanism underlying subtyping is the competition between "group" units and "individual" units, such that stereotype-inconsistent information will be captured by individual units, provided the inconsistencies are concentrated in specific individuals. When the inconsistencies are dispersed, the group units take responsibility for them, updating the group stereotype accordingly. Another finding, also supported by connectionist modeling,[90] is that individuals with moderate inconsistencies cause more updating than individuals with extreme inconsistencies. The logic is once again that extreme inconsistencies cause the individual to be segregated from the group stereotype.

Extinction learning

Like stereotypes, associative memories—in particular, fear memories—are difficult to extinguish once formed. For example, in a typical fear-conditioning paradigm, a rat is exposed to repeated tone-shock pairings; after only a few pairings, the rat will reliably freeze in response to the tone, indicating its anticipation of an upcoming shock. It may take dozens of tone-alone pairings to return the animal to its pre-conditioning response to the tone, indicating that extinction is much slower than acquisition. Importantly, the fact that the rat has returned to baseline does not mean that it has unlearned the fear memory. Under appropriate conditions, the rat's fear memory will return.[91] For example, simply waiting a month before testing the rat's response to the tone is sufficient to reveal the dormant fear, a phenomenon known as *spontaneous recovery*.[92]

As with stereotype updating, one possibility is that conditioned fear is resistant to inconsistent information presented during extinction because the extinction trials are regarded as outliers or possibly subtypes. Thus, although fear can be temporarily reduced during extinction, it is not erased because the subtyping process effectively immunizes the fear memory from disconfirmation. In support of this view, there are suggestive parallels with stereotype updating. Analogous to the dispersed inconsistency conditions in stereotype updating studies,[93] performing extinction in multiple contexts reduces the return of fear.[94] Analogous to the moderate versus extreme inconsistency manipulation,[95] gradually reducing the frequency of tone-shock pairs during extinction prevents the return of fear,[96] possibly by titrating the size of the error signal driving memory updating. More generally, it has been argued that effective memory-updating procedures must control the magnitude of inconsistency between observations and the memory-based expectation, in order to prevent new memories from being formed to accommodate the inconsistent information.[97]

Conspiracy theories

According to the legal scholars Cass Sunstein and Adrian Vermeule, a conspiracy theory is "an effort to explain some event or practice by reference to the machinations of powerful people, who attempt to conceal their role (at least until their aims are accomplished)."[98] Conspiracy theories are interesting from the perspective of auxiliary hypotheses because they often require a spiraling proliferation of auxiliaries to stay afloat. Each tenuous hypothesis needs an additional tenuous hypothesis to lend it plausibility, which in turn needs more tenuous hypotheses, until the theory embraces an enormous explanatory scope. For example, people who believe that the Holocaust was a hoax need to explain why the population of European Jews declined by 6 million during World War II; if they claim that the Jews immigrated to Israel and other countries, then they need to explain the discrepancy with immigration statistics, and if they claim that these statistics are false, then they need to explain why they were falsified, and so on.

Because conspiracy theories tend to have an elaborate support structure of auxiliary hypotheses, disconfirming evidence can be effectively explained away, commonly by undermining the validity of the evidence source. As Sunstein and Vermeule put it:

> Conspiracy theories often attribute extraordinary powers to certain agents—to plan, to control others, to maintain secrets, and so forth. Those who believe that those agents have such powers are especially unlikely to give respectful attention to debunkers, who may, after all, be agents or dupes of those who are responsible for the conspiracy in the first instance . . . The most direct governmental technique for dispelling false (and also harmful) beliefs—providing credible public information—does not work, in any straightforward way, for conspiracy theories. This extra resistance to correction through simple techniques is what makes conspiracy theories distinctively worrisome.[99]

This description conforms to the Bayesian theory's prediction that a sparse, deterministic set of ad hoc auxiliary hypotheses can serve to explain away disconfirming data. In particular, conspiracy theorists use a large set of auxiliary hypotheses that perfectly (i.e., deterministically) predict the observed data and only the observed data (sparsity). This "drive for sense-making"[100] is rational if the predictive power of a conspiracy theory outweighs the penalty for theory complexity—the Bayesian "Occam's razor."[101]

Some evidence suggests that the tendency to endorse conspiracy theories is a personality trait or cognitive style: people who endorse one conspiracy theory tend to also endorse other conspiracy theories.[102] One possibility is that this reflects parametric differences in probabilistic assumptions across individuals, such that people with very sparse and deterministic priors will be more likely to find conspiracy theories plausible.

Religious belief

While conspiracy theories are promulgated by relatively small groups of people, religious beliefs are shared by massive groups of people. Yet most of these people have little or no direct evidence for God: few have witnessed a miracle, spoken to God, or wrestled with an angel in their dreams. In fact, considerable evidence, at least on the surface, argues against belief in God, such as the existence of evil and the historical inaccuracy of the Bible.

One of the fundamental problems in the philosophy of religion is to understand the epistemological basis for religious beliefs—are they justified,[103] or are they fictions created by psychological biases and cultural practices?[104] Central to this debate is the status of evidence for the existence of God, such as reports of miracles. Following Hume (1748), a miracle is conventionally defined as "a transgression of a law of nature by a particular volition of the Deity" (p. 173). Hume famously argued that the evidence for miracles will always be outweighed by the evidence against them, since miracles are one-time transgressions of "natural" laws that have been established on the basis of countless observations. It would require unshakeable faith in the testimony of witnesses to believe in miracles, whereas in fact (Hume argues) testimony typically originates among uneducated, ignorant people.

As a number of philosophers have pointed out,[105] Hume's argument is weakened when one considers miracles through the lens of probability. Even if the reliability of individual witnesses was low, a sufficiently large number of such witnesses should provide strong evidence for a miracle. Likewise, our beliefs about natural laws are based on a finite amount of evidence, possibly from sources of varying reliability, and hence are subject to the same probabilistic considerations. Whether or not the probabilistic analysis supports the existence of God depends on the amount and quality of evidence (from both experiment and hearsay) relative to the prior. Indeed, the same analysis has been used to deny the existence of God.[106]

The probabilistic analysis of miracles provides another example of auxiliary hypotheses in action. The evidential impact of alleged miracles depends on auxiliary hypotheses about the reliability of testimony. If one is a religious believer, one can discount the debunking of miracles by questioning the evidence for natural laws. For example, some creationists argue that the fossil record is fake. Conversely, a non-believer can discount the evidence for miracles by questioning the eyewitness testimony, as Hume did. One retort to this view is that symmetry is misleading: the reliability of scientific evidence is much stronger than the historical testimony (e.g., Biblical sources). However, if one has a strong a priori belief in an omnipotent and frequently inscrutable God, then it may appear more plausible that apparent disconfirmations are simply examples of this inscrutability. In other words, if one believes in intelligent design, then scientific evidence that contradicts religious sources may be interpreted as evidence for our ignorance of the true design.[107]

Conceptual change in childhood

Children undergo dramatic restructuring of their knowledge during development, inspiring analogies with conceptual change in science.[108] According to this "child-as-scientist" analogy, children engage in many of the same epistemic practices as scientists: probabilistically weighing evidence for different theories, balancing simplicity and fit, inferring causal relationships, carrying out experiments. If this analogy holds, then we should expect to see signs of resistance to disconfirmation early in development. In particular, the psychologists Alison Gopnik and Henry Wellman have argued that children form ad hoc auxiliary hypotheses to reason about anomalous data until they can discover more coherent alternative theories.[109]

For example, upon being told that the earth is round, some children preserve their pre-instructional belief that the earth is flat by inferring that the earth is disc-shaped.[110] After being shown two blocks of different weights hitting the ground at the same time when dropped from the same height, some middle-school students inferred that they hit the ground at different times but the difference was too small to observe, or that the blocks were in fact (contrary to the teacher's claims) the same weight.[111] Children who hold a geometric-center theory of balancing believe that blocks must be balanced in the middle; when faced with the failure of this theory applied to uneven blocks, children declare that the uneven blocks are impossible to balance.[112]

Experimental work by Laura Schulz and her collaborators has illuminated the role played by auxiliary hypotheses in children's causal learning.[113] In these experiments, children viewed contact interactions between various blocks, resulting in particular outcomes (e.g., a train noise or a siren noise). Children then made inferences about novel blocks based on ambiguous evidence. The data suggest that children infer abstract laws that describe causal relations between classes of blocks.[114] Schulz and colleagues argue for a connection between the rapid learning abilities of children (supported by abstract causal theories) and resistance to disconfirmation: the explanatory scope of abstract causal laws confers a strong inductive bias that enables learning from small amounts of data, and this same inductive bias confers robustness in the face of anomalous data by assigning responsibility to auxiliary hypotheses (e.g., hidden causes). A single anomaly will typically be insufficient to disconfirm an abstract causal theory that explains a wide range of data.

The use of auxiliary hypotheses has important implications for education. In their survey of the educational literature on this topic, Clark Chinn and William Brewer point out that anomalous data are often used in the classroom to spur conceptual change, yet "the use of anomalous data is no panacea. Science students frequently react to anomalous data by discounting the data in some way, thus preserving their preinstructional theories."[115] They provide examples of children employing a variety of discounting strategies, such as ignoring anomalous data, excluding it from the domain of the theory, holding it in abeyance (promising to deal with it later), and reinterpreting it. Careful attention to these strategies leads to pedagogical approaches that more effectively produce theory change. For example, Chinn and Brewer recommend helping children construct necessary background knowledge before introduction of the anomalous data, combined with the presentation of an intelligible and plausible alternative theory. In addition, bolstering the credibility of the anomalous data, avoiding ambiguities, and using multiple lines of evidence can be effective at producing theory change.

Is the Bayesian analysis falsifiable?

The previous sections have illustrated the impressive scope of the Bayesian analysis, but is it too impressive? Could it explain anything if we're creative enough at devising priors and auxiliaries that conform to the model's predictions? In other words, are Bayesians falling victim to their own Duhem-Quine thesis? Some psychologists say yes—that the success or failure of Bayesian models of cognition hinges on ad hoc choices of priors and likelihoods that conveniently fit the data.[116]

It is true that Bayesian models can be abused in this way, and perhaps sometimes they are. Nonetheless, Bayesian models *are* falsifiable, because their key predictions are not particular beliefs but particular regularities in belief updating. If I can independently measure (or experimentally impose) your prior and likelihood, then Bayes' rule dictates one and only one posterior. If this posterior does not conform to Bayes' rule, then the model has been falsified. Many tests of this sort have been carried out, with the typical result being that posterior judgments utilize both the prior and the likelihood, but do not precisely follow Bayes' rule.[117] The point here is not to establish whether people carry out exact Bayesian inference (they almost surely do not[118]), but rather to show that they are not completely arbitrary.

As this chapter has emphasized, theories consist of multiple hypotheses (some central, some auxiliary) that work in concert to produce observations. Falsification of theories rests upon isolation and evaluation of these individual components; the theory as a whole cannot be directly falsified.[119] The same is true for the Bayesian analysis of auxiliary hypotheses. In order to test this account, we would first need to independently establish the hypothesis space, the likelihood, and the prior. A systematic study of this sort has yet to be undertaken.

Summary

No one likes being wrong, but most of us believe that we *can* be wrong—that we would revise our beliefs when confronted with compelling disconfirmatory evidence. We conventionally think of our priors as inductive biases that may eventually be relinquished as we observe more data. However, priors also color our interpretation of data, determining how their evidential impact should be distributed across the web of beliefs. Certain kinds of probabilistic assumptions about the world lead one's beliefs (under perfect rationality) to be remarkably resistant to disconfirmation, in some cases even transmuting disconfirmation into confirmation. This should not be interpreted as an argument that people are perfectly rational, only that many aspects of their behavior that seem irrational on the surface may in fact be compatible with rationality when understood in terms of reasoning about auxiliary hypotheses.

An important implication is that if we want to change the beliefs of others, we need to attend to the structure of their belief systems rather than (or in addition to) the errors in their belief-updating mechanisms. Rhetorical tactics such as exposing hypocrisies, staging interventions, declaiming righteous truths, and launching informational assaults against another person's central hypotheses are all doomed to be relatively ineffective from the point of view articulated here. To effectively persuade, one must incrementally chip away at the "protective belt" of auxiliary hypotheses until the central hypothesis can be wrested loose. The inherent laboriousness of this tactic may be why social and scientific progress is so slow, even with the most expert of persuasion artists.

7

Seeing patterns

Let thine eyes be thy cook.

—WILLIAM SHAKESPEARE

On December 17, 1996, the Seminole Finance Company building in Clearwater, Florida, was the site of an unusual discovery: a stain, 60 feet tall and 20 feet wide, that appeared to depict the Virgin Mary (Figure 7.1). Hundreds of thousands of visitors traveled to witness the stain in the following weeks, eventually numbering in the millions. A local chemist hypothesized that the stain was produced by weathering combined with water deposits, but this didn't stop a Catholic ministry organization from buying the building, renaming it Our Lady of Clearwater, and installing a rosary factory there.

Examples of religious images appearing in various objects abound. The Virgin Mary has also been discerned in a column of chocolate drippings, a tree stump, a pizza pan, a grilled cheese sandwich, and a fence. The Cone Nebula is sometimes referred to as the "Jesus Nebula" because of its supposed resemblance to Jesus' face. The word "Allah" has appeared on objects ranging from a chicken egg to a satellite image of a tsunami. The Hindu monkey deity Hanuman was reported to be visible on a tree callus in Singapore.

The tendency to perceive patterns even where none exist has sometimes been held up as a distinctive irrationality of human behavior, an overextension of the pattern-discovery machinery that has made our species so successful.[1] In this chapter, I examine these apparently irrational phenomena from a rational inductive perspective. Faced with ambiguity, what kinds of patterns would a rational learner discover? What kinds of inductive biases do people bring to this problem?

FIGURE 7.1. An apparent image of the Virgin Mary on the glass facade of a building in Clearwater, Florida. Source: Polihale, CC BY-SA 3.0.

Coincidences

The belief that some individuals have psychic powers is ancient and ubiquitous. In modern times, television psychics have captivated audiences with their apparent ability to divine hidden truths. The idea of psychic powers has been considered sufficiently plausible by some scientists to merit experimental research. Yet hard evidence for psychic powers is scant. The popular television psychic Sylvia Brown, who claimed an accuracy rate of between 87 and 90%, in fact had a confirmable accuracy of 0%,[2] and many other psychics have been similarly debunked. The scientific efforts to find evidence for psychic powers have been plagued by poor statistical techniques and non-reproducibility.[3]

Why do so many people believe in psychic powers despite the lack of compelling evidence? One answer is that the evidence is typically ambiguous; if you have a disposition to believe in psychic powers, then it will be easy to accept ambiguous evidence in its favor. This is not necessarily irrational, however. In the rest of this section, I'll explore how rational belief formation demarcates the boundary between coincidence and discovery.

Consider the following scenario:

A group of scientists investigating paranormal phenomena have conducted a series of experiments testing people who claim to possess psychic powers. All of these people say that they have psychokinetic abilities: they believe that they can influence the outcome of a coin toss. The scientists tested this claim by flipping a fair coin 100 times in front of each person as they focus their psychic energies.

Even if I told you that the coin landed heads 70 out of 100 times, you'd likely still be inclined to regard this as a coincidence. After all, even a fair coin can sometimes produce a set of improbable outcomes. But after 80 or 90 heads, you might start to suspect that something is going on—is this really a fair coin, or do these people really have psychic powers?

Now consider an alternative scenario:

> A group of scientists investigating genetic engineering have conducted a series of experiments testing drugs that influence the development of rat fetuses. All of these drugs are supposed to affect the sex chromosome: they are intended to affect whether rats are born male or female. The scientists tested this claim by producing 100 baby rats from mothers treated with the drugs.

If you observed that 70 out of 100 babies were born male, you'd likely be inclined to regard this as evidence in favor of the drugs' efficacy. After 80 or 90 male babies, you'd be firmly convinced. Notice that the structure of the drug scenario is identical to the structure of the psychokinesis scenario. The difference in your intuitions stems from the fact that the effect of drugs on genes is a priori plausible, whereas the effect of psychokinesis on coin flips is not. The psychologists Tom Grifiths and Josh Tenenbaum conducted this experiment with college students and found results consistent with these intuitions.[4]

Griffiths and Tenenbaum argued that people use probabilistic inference to determine whether a set of observations is a mere coincidence or a meaningful discovery. The key idea is that a meaningful discovery occurs when observations are surprising under one's current theory, but substantially less surprising under an alternative theory. In contrast, a coincidence occurs when observations are surprising under one's current theory, and no alternative theory is available that makes them substantially less surprising. The distinction between coincidence and discovery is continuous: observations could lie somewhere in between the two poles.

Applying this idea to the coins/genes examples, let's consider two plausible hypotheses suggested by the experimental setup. Hypothesis h_0 (the default hypothesis) states that the binary outcome is fair (50/50 chances) and uninfluenced by psychokinesis or drugs (as the case may be). Hypothesis h_1 (the alternative hypothesis) states that the binary outcome is influenced by the intervention, such that the probability of heads (for coin flips) or males (for drug testing) is ω, which itself is drawn from a uniform distribution. The only difference between these examples is that the prior probability for

h_1 is (for most people) higher in the drug testing example than in the coin flipping example, since the effectiveness of psychokinesis is unproven but the effectiveness of drugs is commonplace. Combining the prior and likelihood according to Bayes' rule, the posterior probability of h_1 is thus generally higher in the drug testing example. Nonetheless, even in the psychokinesis example, people will shift their beliefs towards h_1 given sufficient evidence, consistent with the fact that the likelihood will eventually overwhelm the prior.

Covariation

Let's now look at a slightly more complex scenario in which the data available to the observer is summarized by a 2x2 contingency table, which shows the co-occurrence frequencies of two binary variables (call them X and Y; see Figure 7.2). The task for the observer is to assess the covariation between the two variables: To what extent is one variable more likely to be present or absent when the other is present rather than absent? For example, variable X might be a medical treatment, and variable Y might be a cure, in which case the observer wishes to assess whether an illness is more or less likely to be cured when the treatment is administered.

		Variable Y (e.g., cure)	
		Present	Absent
Variable X (e.g., treatment)	Present	A	B
	Absent	C	D

FIGURE 7.2. An example 2×2 contingency table for two binary variables (X and Y). The letter in each cell denotes the number of times X was present or absent when Y was present or absent.

Here, in contrast to the previous examples, it is important to distinguish the presence/absence of variables (whereas coin flips and gender do not have an intrinsic "on/off" characterization). The reason this distinction is important psychologically is because people do not treat all cells of the contingency table equivalently when assessing contingency. In particular, judgments are more influenced by a variable's presence than by its absence. This asymmetry has been regarded by some psychologists as an irrational quirk of covariation assessment.[5]

It turns out, however, that differential influence is rational under the assumption that the presence of variables is rare; this is the "rarity assumption" that we've already encountered (and justified) in previous chapters.[6] In fact, this can be viewed as another manifestation of the principles underlying the positive test heuristic discussed in Chapter 5. A rare outcome is more informative than a common outcome. Observing a treatment that cures a disease provides more information about the relationship between the two variables than observing the absence of disease in the absence of a treatment, to the extent that the treatment and disease are rare. Consistent with this theoretical analysis, the psychologists Craig McKenzie and Laurie Mikkelsen have shown that one can actually reverse the asymmetry if the presence of variables is made more common than their absence.[7]

Another apparent quirk of covariation assessment is the bias to report positive covariation, even when the variables do not reliably covary.[8] For example, when shown pictures supposedly drawn by patients, along with randomly assigned patient symptoms, people asserted that the picture characteristics covaried with the symptoms.[9] From a Bayesian perspective, however, employing prior knowledge is rational. Most experimental participants expect that the stimuli they're shown are not completely random, and in fact when the possibility of complete randomness is legitimized within the context of the experiment (e.g., simply mentioning randomness as a possible scenario), participants are more accurate at detecting non-contingency.[10] This echoes the finding by Tenenbaum and Griffiths that prior beliefs about causal structure can affect the threshold that people use to adjudicate between mere coincidence and meaningful discovery.

The gambler's and hot-hand fallacies

It was a seemingly ordinary summer night in 1913 at the Monte Carlo Casino, when it was noticed that the roulette ball had repeatedly fallen on black. With each repetition, the expectation grew that the ball must fall on red at the next spin. Finally, after 26 spins, the ball did fall on red, but not before gamblers had lost millions betting on red. These gamblers had committed what is now known as the *gambler's fallacy*: the mistaken belief that random sequences exhibit systematic reversals. While it seems intuitive that a streak must eventually reverse, a little thought reveals why it is a fallacy. There is no way for the roulette ball to "know" how many times it previously landed on black or red; all the outcomes are identically and independently distributed. This fallacy has been observed in other domains. For example, when people buy lottery tickets, they are less likely to buy a particular number after it's been drawn.[11] In soccer penalty shoot-outs, after a sequence of kicks in the same direction, goalkeepers are more likely to dive in the opposite direction.[12] A similar tendency to switch after streaks has been observed in the decisions of refugee asylum court judges, loan application reviews, and professional baseball umpire pitch calls.[13]

The puzzle of why the gambler's fallacy occurs will be addressed shortly, but first I need to describe another layer of the puzzle. Sometimes people exhibit an apparently opposite bias: the prediction that a streak will continue. For example, observing a basketball player who scores several shots in a row, observers (and basketball players themselves!) expect that the player has a "hot hand" that empowers him to continue the streak.[14] Indeed, even traders in betting markets, who have a strong financial incentive to be accurate, appear to believe in hot hands.[15] Each of these fallacies is puzzling by itself, but even more puzzling is how they could possibly coexist.

The economists Matthew Rabin and Dimitri Vayanos have developed a theory to address this puzzle.[16] They consider observers who update their beliefs rationally but have possibly mistaken assumptions about the structure of the environment. In particular, observers believe that the signals they observe (e.g., a basketball player's performance) depend on a hidden state (e.g., the player's ability) plus some random noise (e.g., luck). The hidden state drifts gradually and randomly over time, tending to revert towards some average ability level. The noise corrupting the hidden state is assumed to be anti-correlated across time (a player's good luck in one game is likely to produce bad luck in the next game). This is the critical assumption underlying the gambler's fallacy. The hot-hand fallacy arises because the hidden state is positively correlated over time (a team with high ability tends to remain high ability). Given these assumptions, a rational belief-updating observer will infer that short streaks are likely to reverse, whereas long streaks are likely to continue. Intuitively, if reversals fail to occur for a long period of time, the observer becomes confident that the signals reflect the underlying hidden state (a player's performance reflects ability rather than fluctuating luck), consistent with experimental findings.[17] This pattern can also be seen in real-world lottery players: as already mentioned above, people are less likely to buy a number if it won the previous week, but more likely to buy the number if it was drawn for several consecutive weeks prior to the previous week.[18]

The Rabin and Vayanos theory provides insight into another factor that affects the balance between the two fallacies. Several experiments have found that when people are convinced that sequences are less random (e.g., by telling them that signals are positively correlated over time), the hot-hand fallacy tends to increase relative to the gambler's fallacy.[19] According to the theory, belief in less randomness implies that streaks should be attributed to the hidden state rather than noise. In other words, low randomness promotes the perception of ability that persists over luck that reverses.

Summary

The ability to detect patterns in ambiguous data is one of the most powerful tools in the arsenal of the human brain. Does this tool run amok, leading us into false beliefs about non-existent patterns? On the one hand, the answer is yes: we do sometimes see patterns where none objectively exist. On the other hand, this is not necessarily irrational. We can use inductive biases rationally to grapple with ambiguity, and these biases have an adaptive function. Fictitious pattern detection is a necessary by-product of having such biases.

8

Are we consistent?

Do I contradict myself? Very well then I contradict myself,
(I am large, I contain multitudes.)

—WALT WHITMAN

Consistency is a pillar of rationality. For example, if I prefer option A to option B, then this preference should hold regardless of what other options are added to the choice set. Even if a new option C is preferred over A and B, I should never reverse my preference for A over B. That, at least, is the standard story in economics. Its theoretical foundation comes from a set of axioms that specify conditions under which we can interpret an agent's behavior as maximizing expected utility (i.e., utility averaged over any randomness in the outcomes; see Chapter 2).[1] While this foundation allowed economists to define and study a precise notion of rationality, it came under attack from psychologists as an empirical description of human behavior.[2] People do not appear to be consistent in the way required by expected utility theory.

However, this is not the whole story. The standard economic analysis that has been applied to these apparent inconsistencies assumed that people are perfect information processors, that they have complete knowledge of market conditions, and that they have complete knowledge of their own internal mental states. Once we relax these assumptions, we will find that inconsistencies in decision making may be compatible with rationality after all.

Decoys

Suppose you are thinking about buying new shoes. One pair of shoes (call it A) is of low quality but cheap, whereas the other pair (call it B) is of high quality but expensive. You're having trouble making up your mind, when you notice another pair (D, the decoy) that is both more expensive and of lower

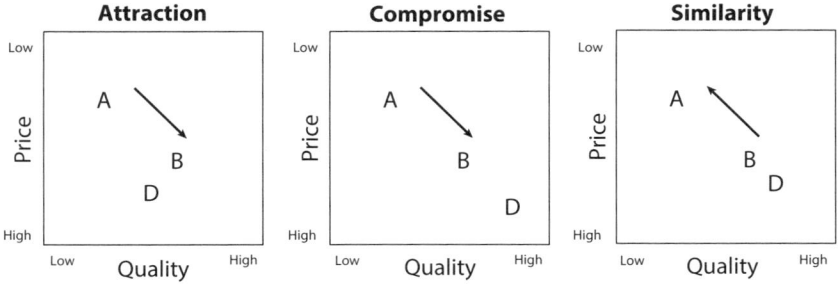

FIGURE 8.1. Decoy effects. Axes measure option attributes (e.g., quality and price). Each letter indicates an option. Option A is cheaper than B but also lower quality. The preference for A relative to B changes as a function of where the decoy D is positioned in the attribute space, as shown by the arrow. For example, the attraction effect occurs when D is close to B in attribute space, but B is more desirable than D along one or both attributes, causing a shift in preference from A towards B.

quality than B (schematized in Figure 8.1, left panel). Obviously you'd never buy D. Nonetheless, D has an *attraction* effect: you become more inclined to choose B.[3] Intuitively, B "looks better" compared to A when contrasted with D. Similarly, B looks better when D is made both very expensive and of very high quality (e.g., designer shoes that are showcased near cheaper brands). Now B looks like a compromise between A and D (Figure 8.1, middle panel).[4] But if you move D closer to B (while keeping it more expensive and of higher quality; Figure 8.1, right panel), then the opposite effect occurs: your choices shift towards A.[5]

To make sense of these effects, let's first go over some basics about multi-attribute decision making. Generally, there exists an "indifference curve" through attribute space (Figure 8.2) that defines combinations of attributes (e.g., price and quality) that yield the same product value. A consumer is by definition indifferent to distinctions between any products that lie on this line. Firms and consumers collectively determine a "fair market value" for a particular product, corresponding to a particular indifference curve. Products that lie above the indifference curve are considered "good deals"—consumers would be willing to pay more for the same quality. Products that lie below the indifference curve are considered "bad deals"—consumers would be unwilling to purchase the product unless its price was reduced or its quality was increased. Thus, a rational firm should design their products to lie on the indifference curve.

The problem, from the consumer's perspective, is that often he doesn't have direct access to the fair market value for products. Consider, again, the problem of buying shoes. You see a set of options with varying price/quality

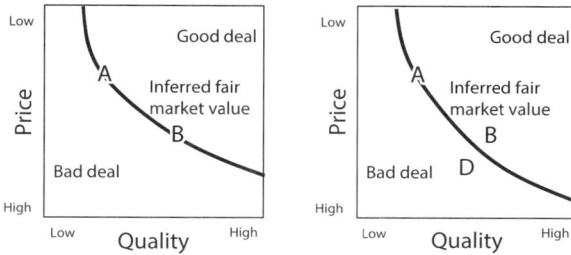

FIGURE 8.2. Illustration of how a decoy (D) changes the inferred fair market value (shown as an indifference curve).

combinations, but you may not know whether a particular price is fair for a particular level of quality. It's reasonable to assume that the shoe store selects its products based on market conditions, so that the shoes on offer are close to the indifference curve. It's also reasonable to assume that the shoe store is catering to consumers with different tastes (i.e., consumers who want different price/quality trade-offs), so the particular price/quality combinations on offer are probably representative of the taste distribution. If most people like intermediate levels of both price and quality, then most shoes will occupy that part of attribute space.

With these assumptions, the savvy consumer can infer something about the fair market value just from the options on offer in the shoe store.[6] Intuitively, the inferred fair market indifference curve should pass nearby all of the options. This means that adding a new option can make other options look better or worse. Figure 8.2 illustrates this for the attraction effect. Adding the decoy D pushes the inferred fair market indifference curve down in the vicinity of option B, making B look like a "good deal."

The same explanation applies to the compromise effect, assuming that attributes tend to be imperfectly substitutable (e.g., people are relatively insensitive to price differences for low-quality products, and relatively insensitive to quality for expensive products). This is why the indifference curves bend near the extremes.

But why does the effect reverse when the decoy D is placed close to B in attribute space, such that now A becomes more desirable (the similarity effect)? The cognitive scientists Pradeep Shenoy and Angela Yu offered the following explanation.[7] If people choose options with probability proportional to their inferred values, then D will effectively compete with B. Critically, because B and D are close to one another, their inferred values will be correlated. Under different market conditions, B and D will tend to either both be better than A or both be worse. Thus, in the presence of D, on average

A will tend to be chosen more often than B, whose probability will be split with D, compared to A being chosen equally often without the decoy (assuming A and B are equally desirable prior to the introduction of the decoy). In Chapter 12, we will revisit the question of why people choose probabilistically rather than deterministically.

To summarize so far, decoy effects are not necessarily irrational violations of consistency. We can make sense of these effects by understanding the inference problem facing consumers. When market conditions are ambiguous, what constitutes a good trade-off between attributes will also be ambiguous. Decoys shift beliefs about market conditions and thus change the perception of these trade-offs.

When more is less

Continuing the shoe-shopping example, imagine you enter a shoe store and see a "buy one, get one free" deal. You find a pair of shoes you'd like to buy, and then see which free shoes are on offer. To your chagrin, you discover that the free shoes on offer are covered with holes and stains. Compare this "deal" option to a "no-deal" option where the desirable shoes aren't paired with the tattered shoes. Which is a better deal? Intuitively, getting free stuff is always better than not getting free stuff, even if the free stuff is low quality. So sure, you'll take the free stuff. You might even be willing to pay a bit extra for the tattered shoes.

What if you only got to evaluate one of the two options? For example, suppose you and your friend go to the same shoe store at different times; when you visit the store, you evaluate the deal, but only the no-deal option is available when your friend visits. The critical question is how your willingness to pay for the deal differs across these joint and separate evaluation scenarios.

The behavioral scientist Christopher Hsee studied this question using a number of different cover stories.[8] For example, he elicited willingness to pay for a dinnerware set that either included or didn't include a set of broken dishes. Hsee found that if separate groups of people evaluate the set with and without broken dishes, they value the set without broken dishes more highly. But this reverses if the same people evaluate both sets at the same time. Similarly, separate groups of people are willing to pay more for a 7-ounce serving of ice cream that overfills a 5-ounce cup than an 8-ounce serving of ice cream that underfills a 10-ounce cup; but when the same people evaluate both options, their preferences reverse, favoring the larger quantity of ice cream. Apparently, people do not consistently evaluate products; they have a different evaluation for the same product depending on what else they are evaluating.

This is not a quirk of laboratory experiments. The economist John List has documented the same "more is less" effect in a real-world marketplace (baseball card bidding).[9] When evaluated separately, bidding prices were lower for a pack of 13 cards, 10 of which were described as being in mint condition, compared to a pack containing only the 10 mint condition cards. This preference reversed when the two packs were evaluated jointly.

In agreement with the earlier discussion of decoy effects, Shlomi Sher and Craig McKenzie have proposed that the "more is less" effect arises from inferences about attribute distributions.[10] The basic idea is that a product's attributes are evaluated relative to the distributions from which they were drawn. Since the consumer does not have direct access to these distributions, they must be inferred, and the choice set itself provides information about these distributions. When the dinnerware sets are evaluated separately, the set with broken dishes seems bad relative to the inferred distribution of dinnerware sets (in particular, expectations about how many intact vs. broken items a typical dinnerware set will have). When the dinnerware sets are evaluated jointly, a common distribution is inferred on the basis of both sets, and the set with broken dishes seems good relative to this distribution.

One point that I've glossed over in this section and the previous section is the assumption that valuation is fundamentally relative with respect to some distribution (about the market or attributes). Why should value be relative? In Chapter 10, I revisit and justify this assumption in terms of more fundamental constraints on how our brains process information.

Constructing preferences

So far, we've considered how preference inconsistency can arise from inferences about the external world (market conditions and attribute distributions). A more radical extension of this idea is to apply it to inferences about the *internal* world of our minds. If we have imperfect knowledge about the causal mechanisms generating our behavior, then we must make inferences about them, much in the same way that we reason about the behavior of other people. Echoing a recurrent theme of this book, these inferences may lead to systematic errors (preference construction), even if the inferential process itself is sound.

Some of the strongest evidence for preference construction comes from studies in which the experimenters manipulated the preferences of their subjects without the subjects knowing.[11] In one famous example, the psychologists Richard Nisbett and Timothy Wilson placed articles of clothing on a table in a public marketplace and told passersby that they were conducting a consumer survey. People who stopped at the table were asked to judge which article of clothing was of the best quality, and then to explain their choice. Unbeknownst to the subjects (but known to Nisbett and Wilson), there was a tendency to choose objects on the right side of the table. This means that the same article of clothing would be judged differently (across subjects) depending on whether it was placed on the left or right side of the table. In the case of stockings, subjects chose them almost four times as often when they were on the right side of the table compared to the left side. Since subjects were unaware of this bias, they never mentioned it in their explanations. In fact, when asked directly about the possible influence of spatial position, almost all the subjects denied the possibility, "usually with a worried glance at the interviewer suggesting that they felt either that they had misunderstood the question or were dealing with a madman."[12]

Another source of evidence for preference construction comes from studies in which the act of choosing alters preferences, such that preferences and choices are in alignment. For example, in the "free-choice paradigm," subjects first rate a variety of items and then make choices between pairs of items that were rated similarly. The key finding is that when subjects are asked to rate the same items again, their ratings of the chosen items increase and their ratings of the unchosen items decrease.[13] It's as though our minds are implicitly reasoning, "I chose that; therefore I must like it." Nonetheless, introspectively the causal arrow feels like it's pointing in the other direction: we feel as though we have stable preferences that produce our choices. This illusion is so strong that we may even misremember our original preferences as being identical to our new preferences.[14] Moreover, our preferences can change simply by professing (falsely) that we like something. In the classic "cognitive dissonance" experiment by Leon Festinger and James Carlsmith, people who were asked to lie that a boring task (turning wooden knobs) was fun subsequently increased their own ratings of how fun the task was.[15]

The illusion of preference stability is vividly illustrated by studies of "choice blindness." These studies were inspired by analogous studies of "change blindness" in visual perception, which showed that people often fail to detect dramatic changes in visual scenes. For example, people can fail to detect that the person they're talking to has changed to a completely different person following a brief interruption, indicating that the brain's visual representation of scenes is much more impoverished than our subjective phenomenology would lead us to believe. In choice blindness studies, the experimenter uses sleight of hand to covertly switch the item chosen by subjects without them noticing. Nonetheless, subjects readily supply explanations for why they "chose" the switched item.[16] Importantly, choice blindness can cause long-lasting preference changes, with people subsequently choosing the switched item, despite the fact that they had initially rejected it.[17]

This even works for political attitudes: people often fail to notice when their statements about political attitudes have been covertly altered to state the opposite attitude; then subsequently they confabulate explanations for why they endorsed the reversed attitude, and this change in attitude is reflected in judgments a week later.[18,19]

The psychologist Fiery Cushman has developed an intriguing account of these and related phenomena.[20] The starting point of this account is that our behavior is determined by multiple distinct "control systems" in the brain, which act in a semi-modular manner (the processes and representations within each system are opaque to the other systems). This means that at a given time, one system might be in control of behavior, and the other systems don't have direct access to the causes of behavior. And even if they did, they wouldn't necessarily be able to understand those causes.

For example, it is well-established that people act on the basis of habits. When I moved to a new house, I would sometimes accidentally drive back to my old house. This seems clearly irrational, a consequence of the fact that one of my control systems executes actions that were previously taken in the past. We can contrast such habitual action selection with a control system based on planning, which uses beliefs about the world (where is my house?) to determine the most efficient sequence of actions that will satisfy my goals (where do I want to go?). Since planning is so effective, why don't we use it all the time? That is, why do we rely on habits at all?

One answer is that while planning is effective, it's not efficient: complex problems might require an unachievable amount of mental effort and time to solve using planning. Suppose I'm deciding whether to accept or reject a job offer. What choice I make now will affect what choices will be available later (e.g., becoming a professor will make it more likely that other professorships will be available to me later). The number of possible future trajectories (sequences of jobs) grows very rapidly, doubling with each extension of the planning horizon. If you think only 10 steps ahead, you will already have to consider over 1000 possible trajectories. And that's just for binary decisions; the situation is much worse if you have to think about more complex decisions, like if you're considering whether to accept or reject two different jobs. Assuming they're mutually exclusive, you would have to consider almost 60,000 trajectories for a 10-step horizon. The point is that planning is hard given the cognitive limitations of people. We can't solve all of our problems by planning; otherwise we would get lost in thought and never do anything. This is why our brains evolved to have different control systems that vary in the computational trade-offs that they make. Habits are a quick and dirty way to make decisions, and they can be highly effective in situations that demand repetitive actions (e.g., making coffee in the morning).

If we take for granted that the brain's planning system can't look arbitrarily far ahead, then it becomes clear that it must consider *sub-goals*—states of the world that are not by themselves rewarding but that signal something about rewards in the future. For example, when planning a vacation, it's not possible to consider all of the future life consequences of the vacation until I die. Rather, I consider a sub-goal like "find a quiet, relaxing place" and then look for plans that satisfy that sub-goal. Because of its limited planning horizon, the planning system doesn't know for sure whether this particular sub-goal will ultimately be good for me (from an evolutionary perspective, what's "good for me" is fitness, my ability to produce offspring). Cushman's key idea is that this sub-goal may have been derived from another control system. The planning system (or another system that sends information to the planning system) observes that I habitually choose quiet, relaxing places for vacations, and let's suppose that these choices were made by the habit system. The planning system can infer that quiet, relaxing places are probably good, since otherwise they wouldn't have become habits. It doesn't know *why* they're good (from the perspective of the habit system). Rather, the planning system constructs new preferences that are consistent with the habitual actions. These preferences (unlike habits) are expressed in the representational format that is comprehensible to the planning system and can now be used in the service of efficient planning. Cushman documents many examples of such *representational exchange*. Behavior, according to this view, is a highly informative conduit for transmitting information between decision-making systems, thus explaining why and how we can learn from observing our own choices.

Intrinsic motivation

In this section, I take the idea of preference construction a step further, analyzing not only what we want, but how much we're willing to work to get it.

The classic economic story about incentives is that people prefer leisure to labor, but will work more if you pay them more. There is a wealth of evidence to support this story.[21] To take one example, the Safelite Glass Corporation, which manufactures glass for automobiles, in 1994 started switching its employees from hourly wages to piece rates (i.e., the employees went from being paid based on how long they worked to being paid based on how much they produced). A subsequent analysis showed that labor productivity increased by 44% per worker following this change.[22] Similarly, mental effort can be increased by offering monetary incentives.[23]

But this turns out to be only part of the story. Incentives can sometimes be demotivating—they have *hidden costs*. In a classic study, the psychologist Edward Deci paid a group of college students to solve a puzzle, while another group of students worked on the puzzle without pay. In a subsequent session, both groups of students worked on new puzzle configurations without pay. Those subjects who were initially paid solved fewer puzzles in this second phase and rated the puzzles as less interesting.[24] This "undermining effect" has been replicated many times with different variations.[25]

The demotivating effect of reward has wide-ranging implications. Paying people a higher piece rate can reduce their performance.[26] Offering compensation for unpopular policy decisions (e.g., building a nuclear power plant) can reduce public support.[27] Paying for blood donations can reduce willingness to donate.[28] Programs designed to incentivize behavior change (e.g., losing weight, wearing seat belts, smoking less) by paying participants might actually backfire. Indeed, paying people consistently produces worse long-term compliance, despite improved short-term compliance.[29] The same logic applies to punishment: imposing fines on late pick-ups from daycare actually increases late pick-ups,[30] and imposing fines on inadequate performance can actually reduce effort.[31]

Psychologists and economists have discussed these phenomena as evidence that extrinsic rewards and punishments "crowd out" intrinsic motivation.[32] One influential explanation of crowding out is that people have uncertainty about how much they like performing certain tasks (possibly because, per Cushman's account, behavior is determined by multiple systems); behavior provides an informative signal about preferences, but this signal can be discounted ("explained away") by alternative causes such as extrinsic incentives.[33] Intuitively, if you were paid money to do something, it's less likely that you were intrinsically motivated, compared to if you did it for free. Similarly, if you were punished for doing something, it's less likely that you were intrinsically motivated not to do it. While this inferential account is appealing given the data discussed in the previous section, the challenge is to explain how incentives can have opposite effects on performance under different circumstances and to identify what those circumstances are.

The economists Roland Bénabou and Jean Tirole propose an answer to this challenge within a decision-theoretic framework. They distill the problem down into the strategic interaction between an *agent* who is performing some task and a *principal* who can intervene to affect the agent's task performance. The agent may have uncertainty about the physical or mental costs of performing the task, and she may additionally have uncertainty about her ability to perform the task. For example, the agent may not know how much effort is required to solve a particular puzzle. The principal has privileged access to some information about the task costs, so the agent can potentially learn about these parameters from the principal's intervention behavior. In particular, the principal can pay a bonus to incentivize performance, and the agent can glean information from this bonus. Intuitively, if the bonus is large, the agent can infer that the task may be costly. As a consequence, the agent will be less inclined to perform the task when the bonus is removed, compared to the case where no bonus was offered at all. Similarly, a bonus can signal that the principal believes the agent has low ability (and thus must be more strongly incentivized to put in greater effort); when the bonus is removed, an agent may abstain from the task due to the inference that the task is too difficult relative to her ability level.

This theory can also be applied internally, in the setting where the agent and principal are the same person. The theory predicts that people may choose to "self-handicap" in order to increase their inferred ability level in the event of unsuccessful performance.[34] Consistent with this idea, the psychologists Steven Berglas and Edward Jones showed that male college students who expected to do poorly on a task opted to take a drug that allegedly inhibited their performance, allowing them to attribute poor performance to the drug rather than to their own ability.[35]

Constructing the self

We construct more than just our preferences and beliefs about cost and ability—we construct entire stories about ourselves that in some ways attain the status of intuitive theories. When we speak of "discovering ourselves," we typically mean that we have changed our intuitive theory in some significant way based on evidence. We may discover, for example, that we are a particular "type" of person (jock, nerd, artist, etc.), much in the same way that we discover the types of other people.[36] The evidence for such discoveries is our own behavior. As discussed above, our imperfect knowledge about the causes of our behavior necessitates that we regard ourselves from the perspective of an external observer: Why did I do that? With preference construction, the answer is relatively superficial (because I like to do that), whereas with self-construction the answer is deeper (because I'm the kind of person who likes to do that).

These ideas can be applied to the study of "moral types." People vary in how prosocial they are—for example, in their disposition towards altruism. If a person has uncertainty about their moral type (e.g., due to imperfect recall of past motives[37]), then they can learn about this hidden variable by observing their own behavior. There is evidence that getting people to act prosocially stimulates them to act more prosocially in the future, consistent with the idea that people infer from their actions that they are a prosocial type. For example, the "foot in the door" paradigm starts by getting people to agree to perform a small prosocial act, which increases the probability that they will subsequently agree to perform a larger prosocial act. In one version of this paradigm, a person is first requested to answer eight questions about consumer goods, and is afterward asked to allow a group of men into their homes for two hours to classify all of the products in their cupboards and closets. About half of the subjects agree to this larger prosocial act after performing the small prosocial act, compared to less than a quarter of the subjects agreeing when the small prosocial request was omitted.[38]

On the other hand, there is also evidence that getting people to act prosocially appears to license antisocial behavior, as though people are collecting karma points through prosocial actions that liberate them in the future to act antisocially. For example, after being given the opportunity to disagree with blatantly sexist statements, people were later more likely to favor a man for a stereotypically male job.[39] Similarly, the opportunity to express a lack of prejudice (e.g., selecting a woman or African American for a job) licensed subsequent prejudicial behavior (e.g., rejecting a woman or African American for a stereotypically male or white job). People will strategically take these licensing actions if they know that they might have to take a morally discrediting action in the future.[40]

We thus have a puzzle: prosocial behavior sometimes promotes future prosocial behavior, and sometimes promotes future antisocial behavior. A key to solving this puzzle is the costliness of prosocial behavior. Only if a behavior is costly to oneself does it serve as a diagnostic signal about your underlying moral type (you wouldn't incur the cost unless you had some intrinsic motivation to be prosocial).[41] Consistent with this idea, the foot-in-the-door paradigm is more successful at inducing the large prosocial act if the small prosocial act is costly.[42] In the original study by Freedman and Fraser, for example, merely asking a person if they'd be willing to answer questions about consumer goods was not as effective as having them actually answer the questions.[43]

A study by Ayelet Gneezy and her colleagues provides a direct demonstration of how costliness influences self-construction.[44] In a "costly" condition, people were given an envelope containing $3 and were told that $2 had been deducted and donated on their behalf to charity. In a "costless" condition, people were given an envelope containing $5 and told that an additional $2 had been donated to charity. Compared to a control condition (just receiving $5), people in the costly condition were subsequently less likely to commit an immoral behavior (lying to potentially take home more money), whereas people in the costless condition were more likely. People in the costly condition also judged themselves to be more helpful and less selfish compared to people in the other conditions. These findings agree with the theory that costly prosocial behavior is an informative signal about one's moral type.

Dynamic inconsistency and multiple selves

Merkel Landis, treasurer of the Carlisle Trust Company, had a great idea in 1909: a "Christmas club" that would help people save money for holiday gifts by allowing them to deposit money into an account that would redisburse the money on December 1st. It was immediately popular, with 350 customers signing up, each saving about $28. Christmas clubs subsequently proliferated and eventually gave way to other "commitment devices" (as they're called in economics). For example, the computer program Freedom will disconnect your internet access for up to eight hours at a time. The company stickK.com allows customers to sign a contract that would force them to forfeit money if they fail to meet a self-determined goal. Commitment devices are ancient: Ulysses ordered his subordinates to tie him to the mast of their ship so that he wouldn't be seduced by the sirens' song, and Han Xin (a famous Chinese general who lived two millennia ago) sent his army into battle with their backs to a river, eliminating the option of escape.

These examples seem natural to us, because we are familiar with the need for self-control. But they are deeply problematic for the theory of rational choice, because they seem to imply *dynamic inconsistency* in our preferences across time. If I really wanted to have money for holiday gifts, why wouldn't I just hold onto it? Putting money into a Christmas club (which historically did not pay interest) is simply a money-losing proposition for a rational agent. This would only make sense if somehow my preferences for money-saving now are different from my preferences in the future. But this challenges the very notion of a unitary rational agent.[45]

One approach to this problem is to concede that we indeed don't have a fixed set of preferences: rather, we consist of a sequence of transient selves, each with its own set of preferences. One self doesn't give a fig for the next self, as the comedian Jerry Seinfeld explains:

> I never get enough sleep. I stay up late at night, cause I'm Night Guy. Night Guy wants to stay up late. "What about getting up after five hours sleep?" Oh, that's Morning Guy's problem. That's not my problem, I'm Night Guy. I stay up as late as I want. So you get up in the morning, you're exhausted, groggy—oh, I hate that Night Guy! See, Night Guy always screws Morning Guy. There's nothing Morning Guy can do. The only thing Morning Guy can do is try and oversleep often enough so that Day Guy looses his job and Night Guy has no money to go out (*Seinfeld*, season 5, episode 2).

The economist Thomas Schelling tells a similar story about his childhood:

> As a boy I saw a movie about Admiral Byrd's Antarctic expedition and was impressed that as a boy he had gone outdoors in shirtsleeves to toughen

himself up against the cold. I resolved to go to bed at night with one blanket too few. That decision to go to bed minus one blanket was made by a warm boy. Another boy awoke cold in the night, too cold to retrieve the blanket and resolving to restore it tomorrow. But the next bedtime it was the warm boy again, dreaming of Antarctica, who got to make the decision. And he always did it again.[46]

Intrapersonal coordination problems of this sort abound. For example, when the delay between ordering and receiving groceries is very short, people will order more "want" items (e.g., desserts) and fewer "should" items (e.g., vegetables), compared to when the delay is long.[47] When the delay is short, the current self recognizes that it can attain higher utility from short-term pleasures, even if this comes at the cost of long-term disutility.

However, this doesn't explain why we would purchase commitment devices. If each self is transient, it should always behave myopically, only optimizing its pleasure during the time period that it exists. What is needed to explain the use of commitment devices is an additional, enduring self—a long-term planner who manages the transient selves. The economists Richard Thaler and Hersh Shefrin developed a formal model based on this idea, which draws an analogy with the principal-agent problem.[48] An agent (e.g., a borrower) makes decisions that affect a principal (e.g., a lender). A problem arises when the agent has incentives that motivate it to act in ways contrary to the interests of the principal. For example, a borrower may take on greater risk after receiving a loan, creating a "moral hazard" for the lender. To mitigate this problem, the lender can try to restrain the borrower's risk-taking behavior (e.g., by imposing credit limits or loan terms). Analogously, Thaler and Shefrin proposed that the long-term self (the planner) can impose rules on the short-term selves (the doers), constraining them to act in the long-term interests of the planner. Commitment devices like Christmas clubs are examples of such rules.

The commitment devices discussed so far are externally enforced contracts. Another important class of commitment devices consists of internally enforced *personal rules*, such as exercise routines, diets, and drug abstention.[49] In a study of taxi drivers, the economist Colin Camerer found that drivers quit for the day when (and only when) they achieve targeted daily earnings, possibly because they use this target as a personal rule to prevent early quitting.[50] What's puzzling about this finding is that taxi drivers adhere to this rule even when it is manifestly disadvantageous. As Roland Bénabou and Jean Tirole put it: "While the self-control value of a daily earnings target seems readily apparent, one must ask what compels the driver, alone and exhausted in his cab, to stick to the rule ex post and stay longer on the job on a bad day when customers are few and far between."[51]

Bénabou and Tirole bring to bear their inferential framework on this question, allowing us to link the idea of multiple selves with the idea of self-inference.[52] As discussed above, people seem to have imperfect recall of their past motives, but are better able to recall their actions, from which they can try to reconstruct their motives. These motives may include transient states (e.g., hunger, fear) as well as more durable character traits. The most important trait, in this context, is willpower: an individual's disposition towards choosing long-term over short-term interests. Because actions are diagnostic of willpower, a person can provide evidence to themselves that their will is strong by taking future-oriented actions, such as strictly adhering to personal rules.

An extreme example of this idea comes from Calvinism, which asserts the doctrine of divine predetermination: "chosen" individuals are at birth assigned by God to Heaven after they die, whereas "non-chosen" individuals are assigned to Hell. The assignment cannot be affected by any earthly actions. One might think, then, that Calvinists would indulge in rampantly heretical behavior. However, the doctrine of divine predetermination also holds that chosen individuals will act virtuously, whereas non-chosen individuals will act sinfully. This means that actions are diagnostically, but not causally, related to one's predetermined status. Thus, a Calvinist can provide self-evidence that they are predetermined for Heaven by acting virtuously.

The psychologists George Quattrone and Amos Tversky devised a clever experimental analog of the Calvinist's situation.[53] People were first asked to submerge their arm in cold water until the pain was intolerable. They were then informed that recent medical studies had discovered a congenital heart condition that could be identified by the effect of exercise on cold tolerance. One group of people in the experiment was told that the heart condition was associated with *increases* in cold tolerance following exercise, whereas another group was told that the heart condition was associated with *decreases* in cold tolerance following exercise. Everyone then rode an exercise bicycle for one minute and repeated the cold tolerance test. The result was that most people exhibited changes in tolerance levels consistent with the absence of the heart condition. Those people who were told that decreased tolerance was diagnostic of a bad heart exhibited greater tolerance, whereas the effect flipped around for the other group. In other words, people produce behavior that is diagnostic of good health, even though it is not causally related to good health.

Is the concept of multiple selves just a metaphor, or is there a sense in which this idea has psychological reality?[54] Studies of intertemporal choice have offered some suggestive evidence for the multiple selves idea, though the conceptualization is somewhat more nuanced than the version developed by economists. When people feel "connected" to their future self, their intertemporal choices are more patient: they are willing to wait longer for a larger reward.[55] Concomitantly, they are less likely to give money to other people when they feel more connected with their future self.[56] This connectedness, although not precisely defined, can be disrupted by life-changing experiences like marriage or starting college. It can also be enhanced—for example, by asking people to write letters to their future self or befriend an avatar representing their future self or by showing them an age-progressed picture of themselves.[57] This enhancement has tangible effects, such as reducing delinquency in children and increasing exercise in adults.

To make sense of these findings, it is useful to think of multiple selves as partially overlapping, with the degree of overlap influenced by external events.[58] If people have uncertainty about the continuity of their selves over time (e.g., "Will I be the same person after I get married?"), then getting people to think about their future selves is a way of self-generating "evidence" that alters their beliefs about self-continuity. The same principle can act retrospectively: when people recall past immoral behavior, they believe that they have changed more compared to when they recall past moral behavior.[59] The evidence available in memory determines inferences about the self across time.

Metapreferences

The idea of self-construction raises an interesting problem: if people make inferences about themselves based on their actions, and thus can in effect choose what kind of self to have, what guides those choices? If preferences can be chosen, then this seems to imply that people have *metapreferences*— preferences over preferences. For example, a person might prefer vegetables over meat, a "first-order" preference that we can write as vegetables $\succ$ meat. A "second-order" metapreference is a preference $A \succ B$ where A and B represent preferences. A vegetarian might not only prefer vegetables to meat, but also prefer that preference over other possible preferences. That is, she might want to be "the kind of person" who prefers vegetables to meat.

The idea of metapreference might seem exotic, but it has played a role in a number of philosophical, economic, and psychological discussions. For example, the philosopher Harry Frankfurt famously argued that metapreferences are an essential ingredient of free will: choosing freely means choosing one's preferences.[60] The economist David George invoked metapreference to analyze how markets can fail to optimize welfare by producing "unpreferred preferences."[61]

If we accept the multiple selves thesis, then a metapreference can be seen as a form of internal commitment device, akin to a personal rule. In his book *Reasons and Persons*, the philosopher Derek Parfit offered the following thought experiment about a Russian nobleman:

> In several years, a young Russian will inherit vast estates. Because he has socialist ideals, he intends, now, to give the land to the peasants. But he knows that in time his ideals may fade. To guard against this possibility, he does two things. He first signs a legal document, which will automatically give away the land, and which can be revoked only with his wife's consent. He then says to his wife: "Promise me that, if I ever change my mind, and ask you to revoke this document, you will not consent." He adds: "I regard my ideals as essential to me. If I lose these ideals, I want you to think that I cease to exist. I want you to regard your husband then, not as me, the man who asks you for this promise, but only as his corrupted later self. Promise me that you would not do what he asks."[62]

The Russian nobleman is not only expressing his preference for socialist ideals, but also his metapreference to maintain this preference in the face of intervening life events.

If we infer preferences on the basis of our choices, can we also infer metapreferences on the basis of our preferences? Returning to the vegetarian example, suppose that she has the opportunity to have dinner at a fancy steakhouse. She might fear that if she tastes an exquisite steak, she will develop a preference for meat, and this would create dissonance between her new preference and her prior metapreference. If the preference provides evidence for inferences about metapreferences, then this will produce a change in her metapreferences. Therefore, she would avoid the steak dinner, despite the fact that it could be quite pleasurable—she opts for self-denial in the service of self-preservation.

To explore these ideas empirically, I ran a study (in collaboration with Tomer Ullman and Laurie Paul) in which I asked people to imagine a pill that could give them new preferences. For example, the pill could make them enjoy receiving shocks (a "masochistic" pill) or make them enjoy delivering shocks to others (a "sadistic" pill). People in the masochistic group read about a hypothetical scenario in which they would be shocked whether or not they chose the pill, and were asked whether they would be willing to pay money for the pill. Similarly, people in the sadistic pill group read about a scenario in which they would be forced to shock another person whether or not they chose the pill, and were asked whether they would be willing to pay money for the pill. Most people (74%) in the masochistic pill group said they would purchase the pill, whereas only 39% of the people in the sadistic pill group said they would purchase the pill. Moreover, people judged that the sadistic pill would induce a bigger change in who they were as a person compared to the masochistic pill. Thus, people are disinclined to change their preferences when they believe it will cause a large change in their self-concept.

People in my study also anticipated that changing their preferences might alter their metapreferences. Most people are willing to purchase a second pill that would reverse the effect of the sadistic pill; moreover, they are willing to purchase a commitment device that would force their future selves to take the second pill. This implies a recognition that their future selves, after taking the first pill, may come to "like" the new preferences (possibly through some form of self-inference), thus creating a misalignment between metapreferences across time, and this misalignment is sufficiently undesirable that people are willing to pay in order to prevent it.

The idea of metapreferences aligning preferences across time may shed light on why moral decisions are sometimes based not on outcomes (as in a utilitarian calculus) but on intrinsic values attached to particular actions (known in philosophy as *deontological* reasoning). For example, most people are unwilling to push someone in front of a trolley in order to save a group of people from an imminent collision, even though they are willing to flip a switch that would divert the trolley from the group and kill the same person.[63] One reason for this divergence of responses could be that people don't want to kill people by physical force, even if it's justifiable from a utilitarian perspective, because this could set in motion a self-inference process whereby they infer that they have a preference for killing others. Indeed, there are some moral scenarios about which people don't even want to *think*. For example, people feel uncomfortable placing a monetary value on marriage and friendship.[64] On this view, thoughts are not harmless—they provide evidence about our preferences, and metapreferences come into play when we seek to regulate the inferential economy of our selves across time.

Summary

Despite our subjective sense that our preferences, beliefs, and identities are internally stable, these properties are in fact remarkably malleable. The sense of consistency derives from the inferences we make based on our own behavior. These inferences are not necessarily irrational. I have argued in this chapter that they reflect the limited information available to people. We do not have direct and centralized access to all the information in our minds— there is no universal search engine into which we can enter queries. Our mind consists of multiple systems for learning and decision making that only exchange limited amounts of information with one another. Even within a single system, there are memory constraints that necessitate inference across time. The end result is that we must constantly construct ourselves. As Timothy Wilson put it, we are "strangers to ourselves."[65]

9

Celestial teapots and flying spaghetti monsters

In formal logic, a contradiction is the signal of defeat, but in the evolution
of real knowledge it marks the first step in progress toward a victory.

—ALFRED NORTH WHITEHEAD

All Niko Alm wanted was to wear a pasta strainer on his head for his driver's
license photo. After waiting three years, and submitting a medical interview
that attested to his mental fitness, he finally received his license from the Austrian police.[1] Several years later, and thousands of miles away, Andrei Filin's
pasta strainer photo was approved by the Main Directorate for Road Safety
Traffic in Moscow, with the warning that if he was ever stopped by the police,
he must be wearing the pasta strainer or else his license will be confiscated.[2]
Dozens of other people around the world have petitioned for the same right,
but the fight continues: in some cases, such as in Ireland and the Nether-
lands, requests to wear pasta strainers in driver's license photos have been
denied.

All these people share a commitment to the recently founded religion
known as Pastafarianism. It all started when the Kansas State Board of Education decided to allow public schools to teach intelligent design as an alternative to evolution. In protest, a young man named Bobby Henderson wrote an
open letter to the board demanding that equal time be given to his Pastafarian
theory of creation, according to which an invisible Flying Spaghetti Monster
created the universe after a heavy drinking bout (the cause of earthly imperfections). Pastafarians believe that evidence for evolution was fabricated by
the Flying Spaghetti Monster as a test of their faith. Its "Noodly Appendage"
undetectably thwarts all scientific efforts to discover the truth.

Henderson's satirical gimmick took on a life of its own, with religious traditions (prayers are concluded with the affirmation "R'amen"), a gospel (which has sold more than 100,000 copies), and even official authorization to ordain marriage in New Zealand. At its core is a simple logical critique of intelligent design: if belief in intelligent design is based on a lack of contradictory evidence, then *any* belief that is not contradicted by the available evidence should be equally valid. But, in fact, all of these beliefs are equally invalid. Advocates of intelligent design, according to this argument, have committed a logical fallacy known as the *argument from ignorance*. The absence of contradictory evidence does not logically entail that a hypothesis is true.

The Pastafarian critique harkens back to a famous thought experiment devised by the philosopher Bertrand Russell:

> If I were to suggest that between the Earth and Mars there is a china teapot revolving about the sun in an elliptical orbit, nobody would be able to disprove my assertion provided I were careful to add that the teapot is too small to be revealed even by our most powerful telescopes. But if I were to go on to say that, since my assertion cannot be disproved, it is intolerable presumption on the part of human reason to doubt it, I should rightly be thought to be talking nonsense. If, however, the existence of such a teapot were affirmed in ancient books, taught as the sacred truth every Sunday, and instilled into the minds of children at school, hesitation to believe in its existence would become a mark of eccentricity and entitle the doubter to the attentions of the psychiatrist in an enlightened age or of the Inquisitor in an earlier time.[3]

Both Henderson and Russell are trying to force the theist's hand, making him accept preposterous assertions or otherwise reckon with the flimsiness of his own beliefs. But Henderson and Russell may have gone too far. The attack is significantly weakened when considered on probabilistic, rather than logical, grounds. Absence of contradictory evidence may not logically entail the existence of God, but it does increase the probability of God's existence. How big of an increase will depend on a number of factors, such as the prior probability of God's existence and the nature of the evidence, as I discuss further below.

The argument from ignorance is an example of an *informal argument*—logically invalid but rhetorically persuasive. My goal in this chapter will be to explain why such arguments are persuasive, by appealing to a Bayesian account first developed by the psychologists Ulrike Hahn and Mike Oaksford.[4] The key idea is that informal arguments are interpreted as providing probabilistic support for hypotheses. As we will see, a number of logical "fallacies" are in fact rational according to the Bayesian account. More importantly, this account explains why some arguments are more or less persuasive as a function of factors (e.g., prior probability) that play no role in a logical analysis.

Bayesian arguments from ignorance

In logical analysis, a conclusion is true given a set of premises if those premises entail the conclusion—that is, if one can derive the conclusion from the premises through the application of inference rules. These rules typically do not depend on the semantics of the premises or conclusion (i.e., what they refer to in the world), but only on their syntax (the roles they play in inferential rules). For example, the *modus ponens* inference rule is illustrated as follows:

> All men are mortal.
> Socrates is a man.
> Therefore Socrates is mortal.

We can express this rule more abstractly by replacing "men," "mortal," and "Socrates" with symbols, so that the inference rule works for any assignment of symbols to meanings (this is what it means for the rule to depend only on syntax):

> X entails Y.
> X is true.
> Therefore Y is true.

The argument from ignorance is a variation of this form:

> X entails Y. ["Evidence contradicts God's existence."]
> X is not true. ["No such evidence has been found."]
> Therefore Y is true. ["Therefore God exists."]

This argument is logically invalid, because such evidence could exist and might simply not have been discovered. Just because it's logically invalid, however, doesn't mean that the argument has no epistemic power. Most people would agree that if your keys are not in one of your pockets, they might plausibly be in the other pocket, even though the latter is not usually logically entailed by the former.

Negative evidence is used ubiquitously by people to update their beliefs. When learning language, children get very few examples of grammatically incorrect sentences; they learn that such sentences are incorrect based almost entirely on generalization from correct sentences.[5] Similarly, a person watching an expert playing a video game can infer that certain objects should be avoided, despite not observing any interactions with those objects.[6] Indeed, it is precisely the absence of such interactions that licenses the inference. The same inference isn't licensed by observing a novice playing the game, because some of those objects might be good and the player simply failed to discover that fact. In both the language and video-game examples, prior knowledge plays a crucial role: what inferences you draw from negative evidence should depend upon your beliefs about how the data were generated.

These intuitions can be formalized using Bayes' rule, which stipulates that the posterior probability of y depends on both its prior probability, $P(y)$, and the likelihood of the negative evidence, $P(\neg x|y)$:

$$P(y|\neg x) = \frac{P(\neg x|y)P(y)}{P(\neg x|y)P(y) + P(\neg x|\neg y)P(\neg y)} \tag{9.1}$$

The Bayesian analysis captures several intuitions about arguments from ignorance. First, logically invalid conclusions can nonetheless get partial credit, provided they have non-zero likelihood and prior. Second, the strength of posterior belief in a particular conclusion is a graded function of prior belief; thus, atheists and theists can reasonably disagree in their beliefs after observing evidence (or lack of evidence). Third, the strength of posterior belief is a graded function of the evidence strength; thus, 100 failed attempts to prove the existence of God will be considered more compelling than 10 failed attempts. All three of these factors have been shown experimentally to influence beliefs in accordance with the Bayesian analysis.[7]

There is another factor that enters into play when we have beliefs about the comprehensiveness of our knowledge. For example, suppose we are consulting a train timetable and find no evidence that the train stops at Hatfield; are we correct in concluding that the train does not stop at Hatfield? The correctness of this conclusion depends on what is known as *epistemic closure*, the assumption that one knows all the conclusions entailed by the available data.[8] Hahn and Oaksford illustrate this point using several variants of the train timetable example:

1. The train does not stop at Hatfield because my friend, who rarely travels on this line, says she cannot recall it ever stopping there.
2. The train does not stop at Hatfield because the railway guard says he cannot recall it ever stopping there.
3. The train does not stop at Hatfield because the railway guard checked the computer timetable, which does not show that it stops there.

Intuitively, the third variant is stronger than the second, which in turn is stronger than the first. On the Bayesian analysis, these differences arise via the likelihood, which encodes the probability that some state of knowledge would obtain given the available data and the procedure for querying the knowledge base. The same person, given the same data, could provide evidence that is weaker or stronger depending on their querying procedure. For example, if the guard is distracted or had just consumed a few drinks at the bar, then we might believe that the querying procedure was non-exhaustive, thereby reducing our confidence that the train does not stop at Hatfield.[9]

Epistemic closure is the basis for another form of argument from ignorance, the "damned by faint praise" phenomenon. Suppose you were reviewing college applications, and you received a recommendation letter stating that the applicant was "polite and punctual." In this setting, epistemic closure requires an assumption about conversational pragmatics, namely that speakers are maximally informative: if the speaker knows something that the listener doesn't know (but would like to know), then the speaker will communicate that information as clearly as possible.[10] According to this principle, you would expect that the letter writer would have more to say about the applicant than an observation about politeness and punctuality, and would therefore say it, unless the additional information was negative. Faint praise is then a signal of missing negative information, despite being an overt expression of positive information.

Adam Harris and his colleagues have explored some of the nuances of the faint praise phenomenon using the Bayesian analysis of arguments from ignorance.[11] If faint praise is preceded by strong praise (e.g., "the applicant is an excellent mathematician"), then missing negative information becomes less likely; if the letter writer had strong negative information, then they would not have reported strong positive information. Thus, according to the Bayesian analysis, the same faint praise can decrease an assessment (when presented on its own) or increase the assessment (when preceded by strong positive information). Harris found that people conformed to these predictions. Moreover, he showed that this only holds true for letters written by a knowledge expert (i.e., someone who knows the applicant well). Since non-experts are less likely to have access to either positive or negative information, their reports do not license the inference of missing information, and therefore faint praise has little effect in either direction.

Circular reasoning

While arguments from ignorance concern the illogically persuasive effect of negative evidence, circular reasoning fallacies concern the illogically persuasive effect of positive evidence. In particular, an argument is circular when the premise already presupposes the conclusion. Hahn and Oaksford give the following examples:

1. God exists because God exists.
2. God exists because the Bible says so, and the Bible is the word of God.

Example 1 is in fact deductively valid, since any proposition entails itself, but most people would not consider it persuasive evidence of God's existence. Example 2, arguably the more common variety of circular argument, is deductively valid if one assumes an implicit premise, namely that the Bible can only be the word of God if God exists.

Hahn and Oaksford argue that the Bayesian approach can shed light on circular arguments with implicit premises. A key distinction is between the observed data (the Bible), which is unambiguous by definition, and the ambiguous interpretation (the Bible is the word of God). Observing the data is indeed informative about the ambiguous interpretation; one could plausibly be convinced that the Bible is the word of God after reading it.

As in the previous applications of Bayesian inference, the degree to which the data are convincing depends on the reader's prior probability and how ambiguous the data are (i.e., the likelihood of the data given the hypothesis). To explore these implications, Hahn and Oaksford presented people with dialogues such as the following:

JOHN: I think there's a thunderstorm.
ANNE: What makes you think that?
JOHN: I just heard a loud noise that could have been thunder.
ANNE: That could have been an airplane.
JOHN: I think it was thunder, because I think it's a thunderstorm.
ANNE: Well, it has been really muggy around here today.

People were asked to judge how convinced Anne should be by John's conclusion that the loud noise was thunder. One group of people were told that "John and Anne are in their trailer home near the airport," making Anne's alternative hypothesis (that the loud noise was an airplane) more likely. Another group of people were told that "John and Anne are in their camper van at their woodland campsite," making the alternative hypothesis less likely. Consistent with the Bayesian analysis, people judged that Anne should be more convinced when the alternative hypothesis was unlikely. While this may seem obvious, note that, based on a logical analysis, the probability of the alternative hypothesis should not have any effect at all; as long as both hypotheses *could* be true, observing thunder does not logically validate either one.

From ad hominem to ad Hitlerum

The radio and television host Glenn Beck is well-known for comparing people he doesn't like to Hitler, and more generally to the Nazis. For example, after Anders Breivik killed 77 people at a Worker's Youth League summer camp in 2011, Beck compared the victims to the Hitler Youth. He also compared the Association of Communities for Reform Now to Hitler's paramilitary group, the Brown Shirts, and he compared the National Endowment for the Arts to Hitler's minister of propaganda, Joesph Goebbels. The comedian Lewis Black commented that "it's like Six Degrees of Kevin Bacon, except there's just one degree, and Kevin Bacon is Hitler!" It's not just Beck who plays this game, however. At the dawn of the internet age, the attorney and author Mike Godwin coined what then became known as Godwin's law: "As an online discussion grows longer, the probability of a comparison involving Nazis or Hitler approaches 1."[12]

There is a technical name for this form of argument: *reductio ad Hitlerum*.[13] It is a specific instance of the ad hominem argument, which uses the character traits of a person to attack that person's conclusions. In particular, the idea is that if a person can be likened to Hitler in some respect, then chances are that the person is like Hitler in other respects that are relevant to the argument at hand. This form of argument is typically considered logically fallacious, because putatively having some trait in common with Hitler does not logically imply that a person has any other traits in common with Hitler. Glenn Beck's hyperbolic statements are examples of how such reasoning can go off the rails.

On the other hand, inferring that a person has some unobserved traits on the basis of observed traits can be a reasonable strategy. If I learn that you enjoy classical music and gourmet French cuisine, I might reasonably infer that you also enjoy ballet. This inference is not *logically* valid, but it is *probabilistically* valid, provided that my assumptions about trait correlations are correct. If people interpret ad hominem arguments in this way, then their degree of belief should vary systematically with the probabilistic evidence provided by the available information. Jaydeep-Singh Bhatia and Mike Oaksford have shown experimentally that people make stronger inferences when an individual's traits are directly relevant to an argument, but still shift their beliefs to some extent even when the traits are not relevant.[14]

People are also sensitive to probabilities when interpreting *ad Hitlerum* arguments.[15] In one study, people were told about good or bad ideas and asked to judge the proportion that Hitler endorsed. For example, "Of all German transportation policies between 1925 and 1945 that historians now recognize as being good, how many do you think Hitler was responsible for?"

This allowed the experimenters to measure the likelihoods, P(Hitler|good) and P(Hitler|bad). They then separately measured judgments about the posterior, P(good|Hitler), for a different set of arguments. For example, after reading the following dialogue, they were asked to judge what person A's opinion should be about a proposed transportation policy:

A: Have you heard about the new transportation policy being considered?
B: Yes, why?
A: I have no idea if it's a good idea or not.
B: It's definitely not?
A: Why?
B: Because Hitler implemented the same policy during his reign.

The study found that the judged goodness of the transportation policy was correlated with the ratio between P(Hitler|good) and P(Hitler|bad), just as predicted by Bayes' rule. In other words, people will reject arguments on the basis of their association with Hitler to the extent that they think Hitler was responsible for bad policies in that domain. People do not seem to be blindly committing a logical fallacy; they are sensitive to the relevant probabilistic quantities and report graded beliefs in accordance with probabilistic reasoning.

Summary

Logical reasoning has long been a favorite punching bag for psychological critiques of rationality. People make systematic mistakes in logical reasoning, which has led some psychologists to posit that judgments are instead based on mental models (imagined possibilities),[16] or on strategies designed to persuade other people rather than to establish the truth values of propositions.[17] The Bayesian approach to informal arguments offers a different viewpoint: the normative standard for such arguments should be probabilistic rather than logical. This viewpoint clarifies why such arguments can be persuasive even when not logically valid.

10

The frugal brain

One sip of reason.

—GEORG LICHTENBERG

Bayesian inference is a solution to the problem of limited data: if we can only observe the world partially, Bayes' rule tells us how to combine our observations with prior knowledge to form beliefs about the hidden parts of the world. But even if our brains had access to unlimited data, they would still be laboring under a set of stringent computational constraints.

To put this in perspective, a conventional personal computer can perform approximately 10 billion basic operations per second, whereas the brain is estimated to perform less than 1000 per second.[1] The precision of these operations is about seven orders of magnitude higher for a personal computer compared to the brain. Moreover, the energy budget of the brain is about an order of magnitude smaller than that of a personal computer (comparable to a dim light bulb), testifying to the astounding energy efficiency of biological computation. Yet even such frugal computation is metabolically extravagant, consuming 20% of the body's energy budget despite only accounting for 2% of the body's mass.

All of this means that our brains must make approximations, which will in turn give rise to systematic errors. We can think about these errors as the result of *approximation biases*, to contrast them with the *inductive biases* that I have focused on so far. Whereas inductive biases are constraints on beliefs about the world, approximation biases are constraints on the process of thought itself. Following the logic of previous chapters, I will ask in this chapter to what extent we can understand errors produced by approximation biases in terms of rational design principles. In other words, are human approximation biases the by-products of useful algorithms?[2]

Efficient coding

When you talk into a cellphone, your voice is encoded into a sequence of binary digits (bits) that is transmitted to the recipient and decoded back into sound. The transformation of an analog signal (like sound waves) into a digital signal (bit sequences) confers a number of advantages. To transmit an analog signal over long distances, one needs amplifiers because the signal dissipates, but amplification adds noise. A digital signal, in contrast, can be transmitted without amplification over long distances and hence is more reliable. In addition, digital transmission is typically faster, uses a higher bandwidth (more information can be transmitted), and requires less power.

Your brain works in a similar way. Your inner ear contains a fluid-filled cavity called the cochlea. When sound waves enter your ear, they cause the fluid to move a tapered membrane that is stiffer at one end, such that different locations along the membrane resonate at different frequencies. These vibrations are converted into electrical signals by tiny hair cells attached to the membrane, ultimately causing neurons in the auditory nerve to fire sequences of *action potentials* (spikes in voltage). Thus, auditory neurons are analog-to-digital converters, encoding sound into bits. All sensory systems in the brain work this way.[3]

Digital communication systems are limited in the number of bits per second they can transmit. Thus, if we can compress messages into digital signals of shorter length, then we can send more of them more quickly. How does compression work? Suppose you call your friend Dave every week, and he always talks about the same thing. If Dave is predictable enough, then the phone conversations will be highly compressible. For example, if Dave simply repeats the word "socks" N times, then naively this message could be transmitted using $N \times K$ bits, where K is the number of bits needed to encode the word "socks." However, this would be very wasteful, because the same information could be transmitted using $M + K$ bits, where M is the number of bits needed to encode the instruction "repeat N times." This coding scheme, by exploiting redundancy in the message, allows a sender to transmit messages of arbitrary length with the same number of bits. Note that even though the transmitted signal is shorter than the original message, the code is *lossless*: the original message can be reliably reconstructed from the signal. In practice (particularly in biological systems like the brain), communication systems are often noisy, and hence the transmitted signal will be *lossy*, meaning that the the original message cannot be reliably reconstructed.

This example illustrates how knowing the structure of messages allows a communication system to work more efficiently by economizing on bits. Information theory, which we already encountered in Chapter 5, formalizes this idea. A message sender is characterized by its probability distribution over messages, $P(m)$, where m denotes a message. A communication system converts the message into a codeword X using the encoding function $x = E(m)$. The codeword is then passed through a (possibly noisy) channel to generate a signal S, which the receiver decodes into an estimate of the original message, $\hat{m}$. Claude Shannon famously proved that the highest achievable bit rate (the number of bits per symbol assuming a constant transmission rate) for near-lossless transmission is the entropy $H(X)$ of the code:[4]

$$H(X) = -\sum_x P(x) \log P(x). \tag{10.1}$$

In other words, the most efficient encoding function should allocate code lengths of $H(X)$ bits on average. This has led to the development of practical compression schemes known collectively as *entropy encoding*, which allocate code length to a codeword proportional to $-\log P(x)$. For example, your computer uses a form of entropy encoding when you save images in JPEG format.

Entropy encoding makes explicit the connection between compression and prediction. Entropy is a measure of a code's unpredictability; more unpredictable messages need longer code lengths. When the entropy is 0, the codeword is perfectly predictable, and hence the sender doesn't even need to send a signal!

When the channel is noisy (Figure 10.1), the achievable bit rate is inevitably lower. More precisely, Shannon proved that the highest bit rate for near-lossless transmission is achieved by choosing the encoding function that maximizes the mutual information $I(X; S)$ between codewords and signals:

$$I(X; S) = H(X) - H(X|S), \tag{10.2}$$

where $H(X|S)$ is the *conditional entropy*, which expresses the degree of uncertainty about the codeword given the signal.[5] When the conditional entropy is 0, the channel is noiseless, and we recover the original upper bound on bit rate.

FIGURE 10.1. Schematic of a communication system.

If you want to increase the bit rate, you need to choose an encoding function that increases the entropy or decreases the conditional entropy. These two desiderata are actually in tension. As we've already seen, increasing the entropy is achieved by eliminating redundancy in the code, thereby making it more compressible. However, this also makes the code less robust to noise. If the noisy channel flips some bits, then codewords can be confused with one another (i.e., the receiver will reconstruct the wrong message), unless there are some redundant bits that can be used to disambiguate the message. In the conversational example given earlier, I could reduce the redundancy of Dave's tedious message by compressing it into the representation "repeat 'socks' N times." However, this compressed representation will be more vulnerable to noise compared to encoding the raw message. If I flip the bits corresponding to one of the repetitions, I can still get a pretty good idea about what the message is about, whereas if I flip the bits corresponding to "repeat," "socks," or "N," then the message will be very hard to reconstruct accurately. I could compromise by keeping around some redundancy—for example, by repeating the compressed message several times. The strategic introduction of redundancy is the basis for the construction of *error-correcting codes*, which are widely used in telecommunication and digital storage.

Efficiency through rank encoding

Now that we have a precise definition of efficient coding, we can discuss how the brain might optimize efficiency. We'll start with a noiseless channel transmitting messages about a one-dimensional magnitude, like loudness or brightness, for a collection of objects. It turns out that one can maximize efficiency (i.e., entropy, assuming a noiseless channel) by coding the object magnitude according to its rank.[6] Simply put, if we sort the magnitudes from smallest to largest and assign to each object a number corresponding to its position in the sorted list of magnitudes, then we can think of the assigned number (the rank) as an object's "rank code." To see how this could work biologically, imagine a neuron that fires spikes in proportion to the number of objects whose magnitudes are less than or equal to a particular magnitude. We can think of each spike as a bit, so that the neuron is outputting a binary codeword over some interval of time.

Some examples are shown in Figure 10.2. Notice that the sensitivity of the rank function to changes in magnitude is greatest for the highest probability magnitudes. This shows intuitively why such a code is efficient: the codewords are most discriminable for magnitudes that are encountered often, so that bits aren't wasted on rare magnitudes.

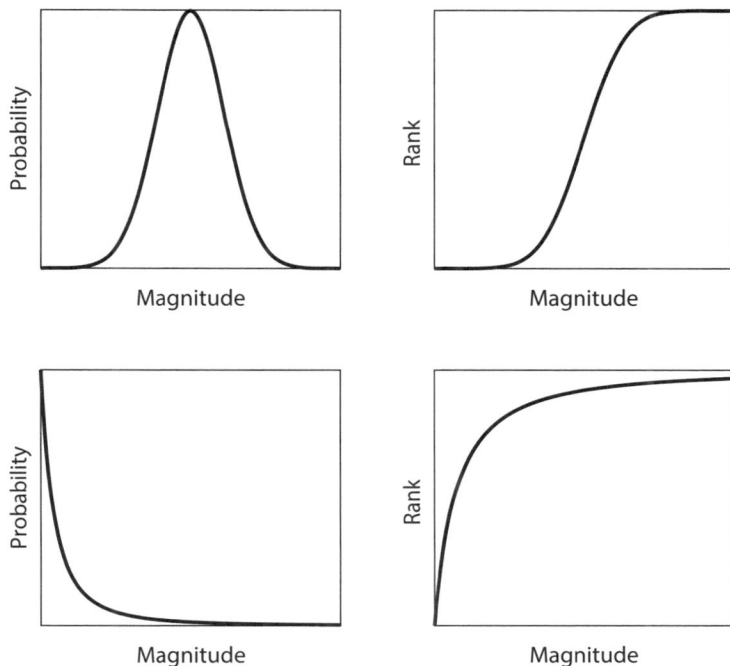

FIGURE 10.2. Examples of magnitude distributions (left) and their corresponding rank encoding functions (right).

A powerful implication of efficient coding is that we can deduce properties of the brain's internal code, and by extension perceptual judgments, from the statistics of the environment. For example, Figure 10.3 (left) shows the log probabilities of different object sizes. This relationship is approximately linear, consistent with the hypothesis that object size m follows a power law:

$$\log P(m) = \alpha_0 - \alpha \log m, \tag{10.3}$$

where α is the slope of the line and α_0 is its vertical offset. Furthermore, the best-fitting value of α is close to 1, which is special because it implies that the rank encoding function is logarithmic: $E(m) \propto \log m$. This is precisely the relationship between objective and subjective size reported in a study by Talia Konkle and Aude Oliva (Figure 10.3, right), who asked people to sort objects (ranging in size from a thumbtack to an airplane) into eight ranks according to their size.[7]

Power-law distributions and logarithmic rank encoding functions are not idiosyncratic to the object size example. Returning to our sound transmission example, it has been shown that the contribution of different frequency

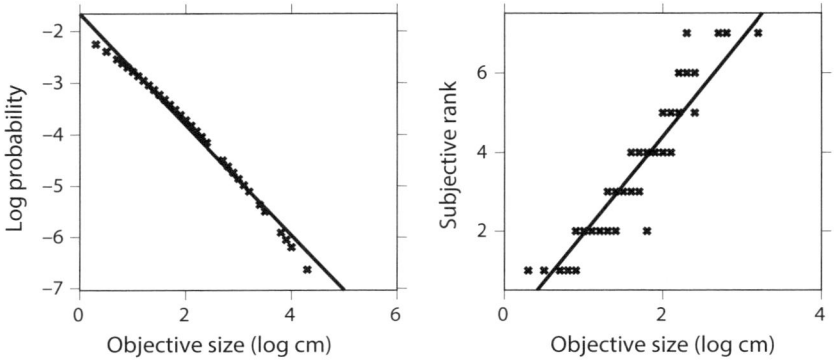

FIGURE 10.3. (Left) The probability distribution of object sizes obeys a power law. (Right) The relationship between objective and subjective size is logarithmic. Data replotted from Konkle and Oliva (2011). Lines show the best-fitting linear relationship.

bands to natural sounds falls off according to a power function of frequency.[8] In other words, natural sounds tend to have relatively little high-frequency content. Consistent with rank encoding, human perception of frequency is approximately logarithmic when loudness is held fixed.[9] Moreover, loudness also follows a power law,[10] and perception of loudness is approximately logarithmic.[11] Audio engineers have long exploited these facts for the purposes of compression by discarding imperceptible sound content.

The logarithmic relationship between objective and subjective magnitude has been observed in numerous sensory modalities since the 19th century. Ernst Weber, one of the founders of experimental psychology, observed that the discriminability of relative magnitude decreased with absolute magnitude. For example, Weber found that the smallest discriminable difference in weight between a test object and a reference object (the "just-noticeable difference" Δm) decreased with the weight (m) of the reference object, such that the ratio $\Delta m/m$ was a constant. Gustav Fechner later showed that this property was consistent with a logarithmic transformation between objective and subjective magnitude,[12] and it is now known as the Weber-Fechner law.

This, however, is not the end of the story. Nearly a century later, the psychologist S. S. Stevens accumulated evidence that the Weber-Fechner "law" fails at very high and low magnitudes.[13] Indeed, this was already apparent in the studies of auditory perception that had found support for a logarithmic relationship at intermediate magnitudes.[14] Stevens argued that this discrepancy could be resolved if one assumed that subjective magnitude instead followed a power law, $E(m) \propto (m - \alpha_0)^\alpha$. At first glance, this looks quite different from the Weber-Fechner law, but if we interpret this as a rank encoding function, then the implied probability distribution is again a power law, but this time the exponent α doesn't need to be equal to 1. Thus, superficially different encoding functions can lead to structurally similar statistical assumptions about magnitudes.

Direct evidence for a power-law encoding function comes from studies that measured the activities of neurons tuned to the magnitude of interest. A typical finding in these studies is that firing rate increases as a power law of stimulus magnitude.[15] Moreover, in some cases the exponent that best fit neural responses was very close to the exponent that best fit perceptual judgments, lending credence to the idea that judgments are linearly related to power law encoding of magnitude in the brain.

Where do power laws come from?

If we locate the origin of logarithmic or power law encoding functions in the ubiquity of power laws, then this raises the question of why power laws are so ubiquitous. There are a number of ways to answer this question, which drive us towards deep ideas in physics and mathematics.[16] I'll briefly discuss two mechanisms for generating power law distributions.

Power law distributions have "heavy tails." What this means, rather counter-intuitively, is that *rare things are common*. Take, for example, one of the most well-known examples of a power law—the frequency of words in natural language.[17] A word like "extraterrestrial" is orders of magnitude rarer than a word like "the," but if you were to select a word at random from a text, you are much more likely to get a rare word than a common word. Thus, to obtain a power law, many low-frequency words need to balance out the small number of high frequency words. Mathematically, this means that the distribution of log frequencies should be close to flat.[18] One generic way to obtain such a flat distribution is if the variable you're measuring (e.g., word frequency) arises from a mixture of other variables (which do not necessarily follow a power law).[19] This is natural in the case of words, where each word belongs to a particular class, such as a part of speech (adjectives, nouns, etc.), so that we can think of the aggregate distribution over word frequency as a weighted average of the class-specific distributions, where the weights correspond to the probabilities of each class. Generally, these classes do not individually follow power laws, but the aggregate distribution does (Figure 10.4).

A quite different path to the power-law distribution for word frequency was developed by Benoit Mandelbrot based on efficient information transmission.[20] Suppose that we had the opportunity to choose word frequencies to maximize the average amount of information per transmission cost. Mandelbrot showed that if the transmission cost is a logarithmic

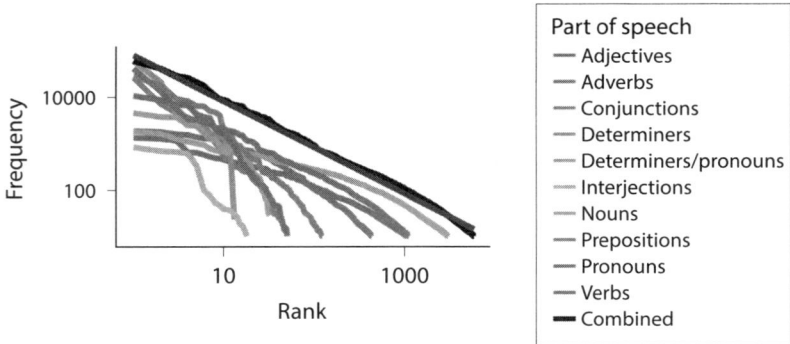

FIGURE 10.4. Word frequency as a function of rank, plotted on a log scale. Dark, straight line shows a power law with a slope of -1. Adapted from Aitchison et al. (2016).

function of a word's frequency rank (i.e., lower frequency words are costlier), then the optimal frequency is a power-law function of rank. This derivation is particularly interesting from a psychological perspective, because it is consistent with the assumption that the brain, if it complies with efficient coding, will allocate longer codewords to rarer words. The next chapter is devoted to further exploring the implications of efficient coding for language.

Local adaptation

So far, I have been treating probability distributions over magnitude as "global" in the sense that they characterize the entire set of magnitudes that one might encounter in one's life. However, greater efficiency could be gained if the brain were able to adapt to local magnitude statistics. Consider, for example, the fact that during the course of a day, the light intensity reaching the retina varies over several orders of magnitude. Light-sensitive neurons in the retina, however, are limited in their range of responses, due to the fact that their firing rates cannot go below 0 and also cannot get arbitrarily large. Helpfully, light intensity at any given time of day stays within a relatively small range of values, since at night it tends to be dark and during the day it tends to be light. Thus, there's no point allocating representational bits to high intensity values during the night, since these values are unlikely to be encountered. To make efficient use of their limited range of firing rates, retinal neurons need only encode the light intensity locally. This is precisely what studies have found: retinal neurons encode light intensity *relative* to the intensity values recently impinging upon those neurons, thereby removing temporal redundancy in the signal.[21] Moreover, these neurons (which are tuned to light

FIGURE 10.5. The Ebbinghaus illusion.

coming from particular spatial locations) also encode light intensity in a spatially local manner by reducing their activity when nearby spatial locations have higher light intensity, consistent with the idea that they are removing spatial redundancy in the signal.

While local adaptation is efficient, it can also give rise to some striking illusions. For example, in the Ebbinghaus illusion (Figure 10.5), the same circle is perceived as smaller when it is surrounded by larger circles (right), and larger when it is surrounded by smaller circles (left). If we think of the flanking circles as providing information about the prior distribution over circle size in this context, then this illusion can be characterized as *repulsion* away from the prior mean. This arises within the efficient coding framework from the fact that increasing the size of the flankers reduces the rank of the central circle.

At this point, recalling earlier chapters, you may be puzzled by the phenomenon of repulsion. Haven't I spent considerable time discussing how the prior should exert an *attractive* effect? According to Bayes' rule, the posterior mean should generally be pulled towards the prior to the degree that the prior is strong and sensory information is unreliable. Indeed, there is evidence for such an attraction effect in judgments of circle size. The psychologists Tim Brady and George Alvarez showed people images of colored circles and, after a delay, asked them to reproduce the size of one particular circle (Figure 10.6; here colors have been converted to shades of gray).[22] They found that the same circle size was attracted towards the average size of other circles with the same color.

There are a number of differences between the experimental procedures that produce the Ebbinghaus illusion and those that produce the attraction effect observed by Brady and Alvarez. In most studies of the Ebbinghaus illusion, people make judgments about circle size while the circles are still visible,

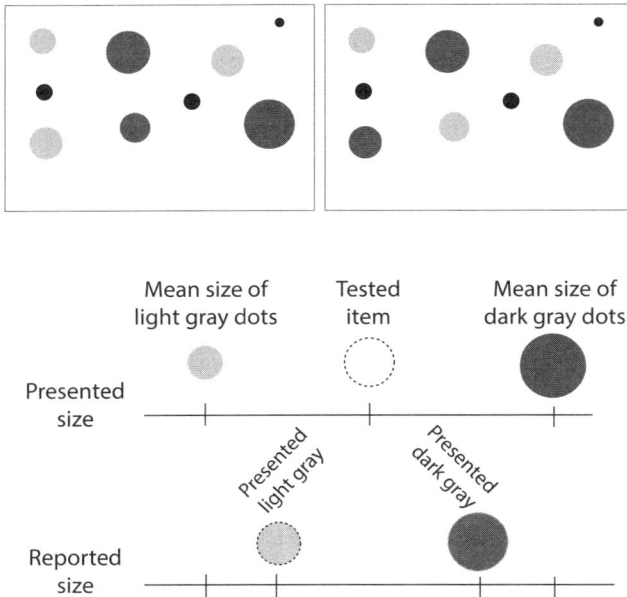

FIGURE 10.6. Size judgments based on memory are attracted towards the average size of circles with the same color (in the original illustration, red and blue; here shown in shades of gray). Adapted from Brady and Alvarez (2011).

whereas in the Brady and Alvarez study, people have to make judgments after the circles have disappeared. However, the Ebbinghaus illusion can be obtained even when size judgments are made based on memory.[23] Another difference is that in the Ebbinghaus illusion, the flankers are arranged in a regular geometric pattern around the central circle, whereas in the Brady and Alvarez study, the circles are randomly arranged. However, the Ebbinghaus illusion can still be obtained when the flankers are randomly arranged and even cluttered together.[24] Thus, these factors cannot explain the differences in results.

The key to resolving the discrepancy between efficient repulsion and Bayesian attraction is to recognize that Bayes' rule enters the efficient coding framework at the decoding stage (see Figure 10.1). In order to make a magnitude judgment, the perceptual system needs to decode the encoded signal back into the original magnitude space. Rank encoding will cause systematic repulsive distortions (pushing the perceived magnitude *away* from the true magnitude) if we assume that the signal has been corrupted by neural noise. In particular, magnitudes with similar ranks will tend to be more confusable with one another. One can show theoretically that repulsion should be stronger when the corrupting noise is large relative to the variability of magnitudes in the context.[25] This happens due to the curvature of the encoding function: magnitudes are more confusable as they get farther away from the prior mean, and therefore the same amount of noise will "spread farther" for more extreme magnitudes, thus enhancing the repulsion effect. Consistent with this hypothesis, repulsion is observed when the contextual magnitudes are similar to the target magnitude (low contextual variability, producing more curvature of the encoding function), or when the noise is larger due to greater memory demands, whereas attraction is observed when target-context similarity and noise are low.[26]

Noisy channels

We have seen in the last section that noise in the mental representation of magnitudes plays an important role in determining the perceptual effects of efficient coding. One way in which noise could arise is if the rank encoding function is approximated by a set of samples.[27] In general, it's reasonable to think that exhaustive enumeration of all magnitudes in a given context (or in the environment as a whole) is intractable for the brain. Even in cases where it might be tractable, it would probably be computationally wasteful, since one can often get a good estimate from a small number of samples[28]—a topic to which we return to in Chapter 12.

Regardless of precisely how noise arises, there is a critical mismatch with our theoretical assumptions so far: recall that the rank encoding function is only information-theoretically optimal when there is no noise. In the presence of noise, it is necessary to introduce some redundancy into the code, in order to reduce the likelihood of decoding the wrong magnitude. One way to do this is to smooth the encoding function, so that similar magnitudes are assigned to similar signals. Although this might seem like it should *increase* decoding errors, it will actually decrease them if the channel is sufficiently noisy, because the error introduced by smoothing will be smaller than the error introduced by noise.[29]

At a psychological level, smoothing could be implemented by reducing discriminability between stimuli.[30] The basic idea is that we construct an approximate rank function by sampling memories of previous magnitudes, and tallying up how many of these retrieved magnitudes are less than the magnitude we are currently judging. If the retrieved magnitude is "blurry" (we can't discern its exact value, but rather a band of plausible values), then nearby magnitudes will get partial credit in the tally.

The optimal level of smoothing (blurring) increases with the level of noise and the variability of magnitudes. Intuitively, when the variability is large, a given magnitude will have fewer "neighbors," and hence more smoothing is necessary to exploit the redundancy between magnitudes. For example, increasing the range of magnitudes will typically increase their variability. As a consequence, one can show that the effect of rank should decrease and another effect should emerge: *range normalization*, whereby a fixed difference in magnitude between two stimuli will be perceived as smaller if the range is larger. Range normalization effects have been observed extensively in perceptual tasks. For example, when people are asked to identify the frequencies of tones within a range, their estimation errors are larger when the range of frequencies is larger,[31] and tones with two different frequencies are more likely to be confused when the range of frequencies is larger.[32]

Our analysis of communication over a noisy channel has led us to two factors (rank and range) that exert distinct effects on perception. These two factors are in tension with each other, since increasing smoothness (a range effect) will inevitably dampen the effect of rank. This tension was recognized many years ago by the psychologist Allen Parducci, who developed a simple descriptive model based on the linear combination of the two factors.[33] The same model can, with a few assumptions, be derived from the information-theoretic framework discussed here, as shown by a postdoc in my lab, Rahul Bhui.[34]

A key advantage of the information-theoretic framework is that it stipulates how the relative weighting of rank and range should be determined (in many of Parducci's applications, the weight is fixed at 0.5). In a categorical judgment task, people assign a stimulus to one of a limited number of categories, where each category corresponds to a bin in the range of magnitudes. A common finding is that people tend to make approximately equal use of all the categories,[35] consistent with rank encoding. However, it has also been found that when the number of categories increases, the effect of rank diminishes, and the effect of range increases.[36] This follows naturally from Bhui's information-theoretic analysis: noise has a more destructive effect when the bins are smaller, so to suppress this noise it is necessary to smooth more aggressively, which corresponds to increasing the weighting of the range component.

Efficient coding beyond perception

The effects of rank and range apply not only to low-level perceptual judgments, like size and loudness, but also to social and economic judgments. Subjective judgments of well-being depend on a person's rank in the wage distribution, as well as on the range of the wage distribution.[37] Rank effects on subjective well-being are at least partly local in space (a person's subjective well-being goes up when their neighbors earn less[38]) and time (subjective well-being goes up when they receive a reward that is greater than their average reward in the recent past[39]). Likewise, judgments of social attributes like competence, hostility, and attractiveness are relative in nature.[40] For example, a face's attractiveness decreases with the range of facial attractiveness present in the local distribution of faces, and increases with the face's attractiveness rank in that distribution.[41] Similar results have been found in judgments of competence,[42] and they even predict real-world behavior like election outcomes.[43]

The fundamentally relative nature of judgment has wide-ranging consequences for economic decision making. The same outcome could be more or less desirable or pleasurable depending on the distribution of possible outcomes. One version of this idea posits that monetary outcomes are evaluated relative to a reference point, which is usually taken to be an individual's current or expected level of wealth.[44] Outcomes greater than the reference point are mentally treated as "gains," and outcomes less than the reference point are mentally treated as "losses." Thus, earning a smaller-than-expected raise, despite being a positive monetary outcome, would be treated as a loss, and getting a smaller-than-expected pay cut, despite being a negative monetary outcome, would be treated as a gain. In laboratory experiments, the same reward is judged to be more pleasurable when it is surprising compared to when it is expected, and surprising losses are more painful than expected losses.[45]

The mental representation of value has three other important properties. First, when the same magnitude of gains and losses (relative to the reference point) are compared, the losses are evaluated more negatively than the corresponding gains are evaluated positively, an effect known as *loss aversion*. Second, sensitivity to the difference in value between two gains diminishes as both gains become larger, a property we have already seen in the encoding of magnitudes by power law or logarithmic functions. Third, sensitivity to the difference in value between two losses diminishes as both losses become larger. Mathematically, we say that the function relating outcomes to values is concave for gains and convex for losses.

Figure 10.7 shows a value function that has these properties, proposed by the psychologists Daniel Kahneman and Amos Tversky.[46] This value function can explain many puzzling phenomena in decision making. For example, the price at which a person is willing to sell a good is typically higher than the price a person is willing to pay for a good, a phenomenon known as the *endowment effect*. In one experiment, when people were given a mug and then offered the chance to sell it, they required as compensation approximately twice what they were willing to pay to acquire the mug.[47] An implication of the endowment effect is that trading volume will appear to be irrationally low,

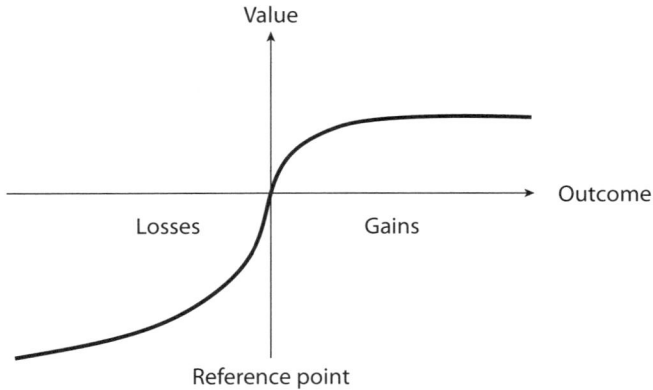

FIGURE 10.7. The value function used in Prospect theory (Kahneman and Tversky, 1979).

violating the Coase theorem, which states that resource allocation in a market should be independent of initial ownership, provided transaction costs are sufficiently low and the allocation of resources does not significantly alter an owner's wealth (as in the case of buying and selling mugs).[48]

Loss aversion can also explain why traders have a tendency to sell assets that have increased in value, while holding assets that have decreased in value (the *disposition effect*).[49] This is arguably an irrational behavior, because there is evidence that some assets, such as stocks, have momentum: stocks that have done well recently are likely to do even better in the future, and stocks that have done poorly recently are likely to do even worse.[50] To explain how the shape of the value function implies the disposition effect, I first need to explain the relationship between the value function and risk preferences. Because outcomes in general may be probabilistic, a standard assumption is that people try to maximize the *expected* value of outcomes, by taking a weighted average, $\sum_x P(x)V(x)$, where $P(x)$ is the probability of outcome x and $V(x)$ is its value. Consider, for example, a choice between $100 for sure or a 50% chance of $200. The expected reward of the gamble is the same as the sure thing, but the expected value will always be lower if the value function is concave. This means that people will be risk-averse for gains (they will tend to avoid risky gambles). Now consider a choice between a loss of $100 for sure or a 50% chance of losing $200. Again, the expected reward of the gamble is the same as the sure thing, but the expected value will always be higher if the value function is convex. This means that people will be risk-seeking for losses. Returning to the case of stock market investment, these properties together imply that people will want to convert risky stocks into non-risky cash when they are increasing in value (risk aversion in the domain of gains) and hold them when they are decreasing in value (risk-seeking in the domain of losses).

A related line of reasoning explains another investment phenomenon known as the *equity premium puzzle*: investment in low-risk assets like bonds far exceeds their market value compared to high-risk assets like stocks.[51] The economists Shlomo Benartzi and Richard Thaler argued that this phenomenon arises from loss aversion.[52] If you invest in a high-risk asset, then you will be confronted with more frequent losses when you check your returns, compared to if you invest in a low-risk asset. If losses are more painful than gains, then you will be driven towards low-risk assets even if this entails a considerable financial opportunity cost. Consistent with this explanation, experiments have shown that giving people fewer opportunities to check their returns or increasing all of the outcomes so as to eliminate losses decreases risk aversion.[53] Note that this does not contradict the previous statement that people are risk-averse for gains and risk-seeking for losses, since people are still risk-averse when the outcomes are increased to eliminate losses; they are just less risk-averse compared to when the gamble can yield both gains and losses, due to the asymmetry of the value function.

While the value function shown in Figure 10.7 can explain the origin of these phenomena, it leaves open *why* the function has this particular form. In fact, we've already seen a number of arguments (the Coase theorem and the existence of stock return momentum) for why one would expect a competitive market to eventually eliminate agents with such value functions. Agents with distorted value functions should be dominated by agents with undistorted value functions, who would be able to efficiently allocate resources and respond to asset momentum. Efficient coding offers an answer that is broadly compatible with the principle of expected utility maximization. If there are constraints on the ability of the brain to accurately represent magnitudes, then the value function will likely be distorted even if the incentive structure of the market militates against such distortions.

According to the efficient coding hypothesis, the value function derives its peculiar shape from the distribution of gains and losses. Specifically, we should be able to reconstruct something that looks like Kahneman and Tversky's value function by plotting the rank function for gains and losses. Neil Stewart and his collaborators did this using bank transaction data as a proxy for real-world gain (credit) and loss (debit) distribution (Figure 10.8).[54] Because small gains and losses are more common than large ones, the rank function exhibits diminishing sensitivity. Note that for losses, the rank function

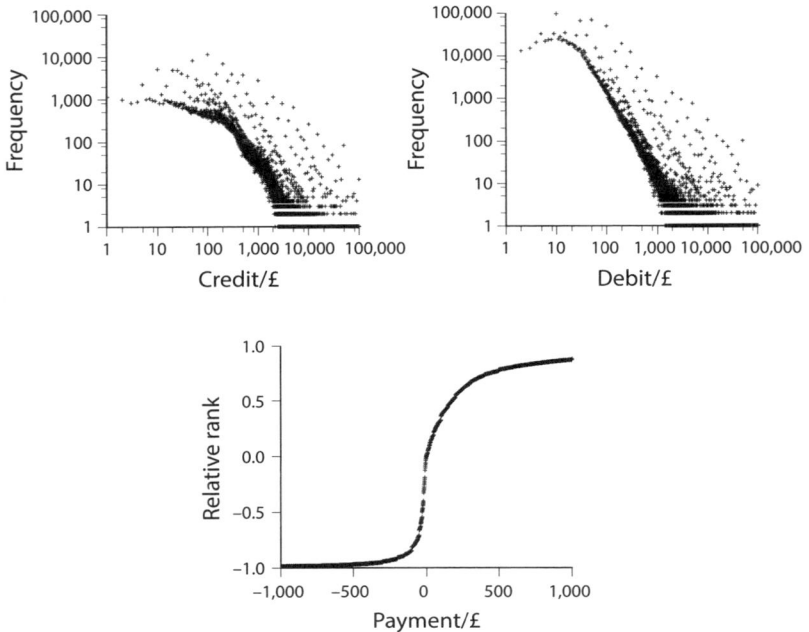

FIGURE 10.8. Probability distribution for credits (top left) and debits (top right). The lower panel shows the value function constructed by concatenating the rank functions for credits (gains) and debits (losses). Adapted from Stewart et al. (2006) and Stewart et al. (2014).

must be flipped upside down to construct the value function (so that larger losses are less valuable), which means that the function is convex in the loss domain. The rank function for losses is also steeper than the rank function for gains, thus capturing the key property that produces loss aversion. This asymmetry can be attributed to the fact that most people are paid in relatively large lump sums, whereas they spend their income in many smaller increments. As a consequence, relatively small losses will have a higher rank than gains of the same magnitude. Stewart showed that a number of other economically meaningful magnitudes, such as reward probabilities and delays, can be analyzed in the same way: mental representations follow rank functions derived from real-world distributions.

Figure 10.8 approximates the global distribution of gains and losses, but people are sensitive to local distributional statistics. For example, the shape of the value function can be manipulated based on the distributions experienced within a single experiment (Figure 10.9).[55] When gains are positively skewed (more small gains than large gains), the rank function is concave, and

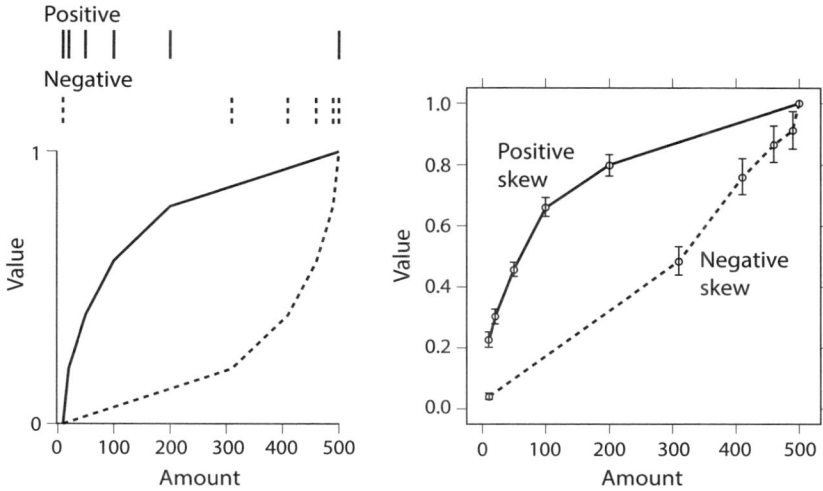

FIGURE 10.9. The effect of positively and negatively skewed reward distributions on the rank function (left) and estimates of the value function from human choices (right). Adapted from Stewart et al. (2014).

people are risk-averse, just as in Kahneman and Tversky's analysis. But when gains are negatively skewed (more large gains than small gains), the rank function is convex, and people are risk-seeking. It has also been shown that loss aversion can be reduced, and even reversed, when the experimental distribution of gains and losses is manipulated so that the rank function for gains is steeper than the rank function for losses.[56] Thus, people do not appear to use a fixed value function, but instead adapt the value function based on local distributional statistics.

Recall that if magnitude encoding is noisy, we expect that magnitude range will exert an effect on the internal representation, by smoothing the rank function.[57] Thus, we expect here (as in the perceptual examples) effects of both rank and range. Consistent with this hypothesis, increasing the range of outcomes has the effect of reducing sensitivity to differences in outcome,[58] and such changes in sensitivity have also been observed in the perception of prices.[59]

Summary

Computation is costly. In this chapter, I have focused on the implications of this cost for encoding and decoding unidimensional magnitudes, ranging from sensory quantities like loudness to cognitive quantities like value. By looking at this problem through the lens of information theory, we can derive hypothetical encoding functions from assumptions about the magnitude distributions and encoding constraints. This led us to a theory of rank and range effects that can explain a remarkably wide range of perceptual and economic phenomena. In the next chapter, I will explore the implications of these ideas for language.

11

Language design

Human speech is like a cracked kettle on which we tap crude rhythms for bears to dance to, while we long to make music that will melt the stars.

—GUSTAVE FLAUBERT

Language is arguably our most powerful, and most unique, tool for the transmission of knowledge. Language is also quite weird. It's filled with ambiguities, non-literal meanings, peculiar syntactic and semantic regularities. Through the ages, these peculiarities have stimulated intrepid attempts to design "better" languages. Yet with a few exceptions (such as Esperanto) the languages inevitably fail to attract a critical mass of speakers.[1] If natural languages are so imperfect, why do they have such stubborn staying power?

In this chapter, I will argue that many peculiarities of natural language are features rather than bugs. They arise from two design principles: informativeness (utterances should convey information about the world) and effort (it should be easy to encode thoughts into utterances and decode utterances back into thoughts).[2] These are, in fact, precisely the principles of efficient coding that I reviewed in the last chapter. Communication channels should be designed such that they communicate as much information as possible. Whereas in the last chapter the code was internal (brain activity) and the communication bottleneck was the limited capacity of neural signaling, in this chapter the code is external (speech) and the communication bottleneck is the limited capacity of speech production and comprehension. The need to balance informativeness and effort has wide-ranging implications for essentially every aspect of language.

Communicative efficiency and ambiguity

The capacity of speech production is defined as the maximal amount of information that speech can convey about thoughts. It is limited by the speed of speech (we can only say a limited number of words per minute) and the lossiness of the encoding function from thought to utterances (the ambiguity of utterances). For example, some words (like "bank" and "fan") have multiple meanings, so a listener can rarely be completely sure which meaning a speaker is referring to when they use those words. Sentences can also have syntactic ambiguity, where different interpretations of the syntactic structure lead to different meanings. "Gary saw Jill with binoculars" could be interpreted to mean Gary used binoculars to see Jill, or that he saw Jill holding binoculars. Syntactic ambiguity is sufficiently common in newspaper headlines that it has its own term—*crash blossoms*, named after the headline "Violinist linked to JAL crash blossoms."[3]

The ambiguity of natural language has always been one of the primary motivations for designers of artificial languages. One of the earliest artificial languages, published by John Wilkins in 1668, sought to remove semantic ambiguity by constructing each word out of a concatenation of semantically unambiguous symbols, each corresponding to a meaningful concept. In recent history, *Loglan* (short for "logical language"), and its successor *Lojban*, were designed to remove as much syntactic ambiguity as possible by using symbolic logic as the basis of their grammar, so that a sentence can only be parsed in one way. The result is extremely cumbersome; the official Lojban grammar handbook concedes that the language designers "tried to err on the side of overkill. There are distinctions possible in this system that no one may care to make in any culture."[4] To express a simple thought requires substantial thinking.

At one gathering of Lojban enthusiasts, guests were served "zalvi ke nakni bakni rectu," which refers to hamburger but translates literally as "ground type of male cow meat."[5] The host of the party (one of the creators of Lojban) was pleased with the outcome, stating, "Mi'a . . . lifri lei xamgu temci," or "Me and others, not including you whom I'm talking to, experienced some particular mass of good time interval."

These examples offer an intuition for why natural language is lossy: removing all ambiguity would require substantial effort from both the speaker and listener. Ambiguity can thus be considered an approximation bias that emerges from the principle of communicative efficiency.[6] This interpretation of ambiguity contrasts with Noam Chomsky's argument that the existence of ambiguity proves that language is not designed for communication:

The natural approach has always been: Is [language] well designed for use, understood typically as use for communication? I think that's the wrong question. The use of language for communication might turn out to be a kind of epiphenomenon . . . If you want to make sure that we never misunderstand one another, for that purpose language is not well designed, because you have such properties as ambiguity. If we want to have the property that the things that we usually would like to say come out short and simple, well, it probably doesn't have that property.[7]

The problem with Chomsky's argument is his assumption that the existence of ambiguity necessitates long and complex utterances. This is not true in general, thanks to the fact that context allows us to disambiguate many otherwise ambiguous utterances. For example, if someone says, "I put the pig into the pen," it's usually safe to assume that they are putting the pig in an enclosure rather than in a writing implement. Logical languages like Lojban try, by construction, to eliminate such ambiguity, thus necessitating redundant communicative effort ("I put the pig into the pen-that-is-an-enclosure"). The principle of effort reduction was aptly summarized by the novelist Elmore Leonard when asked about his secret to writing: "I leave out the parts that people skip."

On the other hand, if one tries to eliminate effort too aggressively, one ends up with situations like the following anecdote that Benjamin Franklin told Thomas Jefferson:

When I was a journeyman printer, one of my companions, an apprentice hatter, having served out his time, was about to open shop for himself. His first concern was to have a handsome sign-board, with a proper inscription. He composed it in these words, "John Thompson, Hatter, makes and sells hats for ready money," with a figure of a hat subjoined; but he thought he would submit it to his friends for their amendments. The first he showed it to thought the word "Hatter" tautologous, because followed by the words "makes hats," which show he was a hatter. It was struck out. The next observed that the word "makes" might as well be omitted, because his customers would not care who made the hats. If good and to their mind, they would buy, by whomsoever made. He struck it out. A third said he thought the words "for ready money" were useless, as it was not the custom of the place to sell on credit. Everyone who purchased expected to pay. They were parted with, and the inscription now stood, "John Thompson sells hats." "Sells hats!" says his next friend. Why nobody will expect you to give them away, what then is the use of that word? It was stricken out, and "hats" followed it, the rather as there was one painted on the board. So the inscription was reduced ultimately to "John Thompson" with the figure of a hat subjoined.[8]

These examples illustrate how informativeness and effort are in tension with each other; an ideal language must balance these two desiderata. The important take-away is that ambiguity is a feature, not a bug, when context provides sufficient information about meanings. If, on the other hand, context is insufficient, then one needs to select utterances that are unambiguous.

We can make the argument about effort more specific if we posit that some linguistic units are costlier to use than others. For example, longer, less predictable, and less frequent words take longer to process during reading.[9] As long as context provides disambiguating information, a linguistic system can reduce effort by mapping multiple meanings to the same less-costly words, rather than mapping the different meanings to distinct words. In other words, less-costly words should be reused as much as possible, and this reuse will result in ambiguity. The psychologist Steve Piantadosi and his collaborators have found quantitative evidence for this hypothesis in the negative relationship between the costliness of a word (as measured by its length, predictability, and frequency) and its reuse (as measured by the number of senses that it has).[10]

The origin of word meanings

Words partition experience into discrete categories, and different languages partition experience in different ways. In light of this cross-linguistic variation, a natural question is whether there are any universal principles governing semantic categorization. One candidate principle is efficient communication: semantic categories should be chosen to balance informativeness and effort.[11]

To illustrate, I will focus on the case of color. Languages differ in the way color words partition the space of colors, using between 2 and 12 words. In their landmark study *Basic Color Terms: Their Universality and Evolution*, the anthropologist Brent Berlin and the linguist Paul Kay argued that languages with different numbers of color terms partition color space in similar ways around a set of "focal" colors (black, white, red, green, yellow, and blue):[12]

- **2 terms**: black and white.
- **3 terms**: + red.
- **4 terms**: + green or yellow.
- **5 terms**: + green and yellow.

- **6 terms**: + blue.
- **7 terms**: + brown.
- **8 or more terms**: + purple, pink, orange, or gray.

Since the Berlin and Kay study, their original universalist position has been challenged by data suggesting that color-naming systems vary across languages in ways that are not adequately predicted by the number of terms alone.[13] For example, the Berinmo language from Papua New Guinea and the Himba language from Namibia both contain five color terms. Although their color terms appear to be organized around the focal colors predicted by Berlin and Kay (black, white, green, red, and yellow), the precise boundaries between colors differ between the languages.[14]

These differences are not mere quirks of how speakers use language, because the boundaries have effects on other aspects of cognition—a family of phenomena known as *categorical perception*.[15] Two colors an equal distance from one another in color space are judged to be less similar if they belong to different categories, compared to if they belong to the same category, and they are less discriminable in memory. Importantly, this is true only for labels defined by a speaker's own language; Himba and Berinmo speakers do not distinguish between the English categories of blue and green, and accordingly do not show the corresponding categorical perception effects for that color boundary. Moreover, the subtle differences in boundaries between Himba and Berinmo produce corresponding categorical perception effects, with patterns of similarity and memory shifting based on the positions of the category boundaries in each language.[16]

The basic question raised by these findings, as well as by the original Berlin and Kay study, is *why* languages partition color space in certain ways and not others. From an efficient communication perspective, the problem is to select a color partition that allows a speaker to communicate as much information as possible about color with the minimum of effort for the speaker and listener (Figure 11.1). The optimal partition depends on the perceptual structure of colors (how easy is it to discriminate between different colors) and communicative need (how often particular colors are referred to). If you want to refer to colors with a limited set of words, you should select categories that (a) allow the listener to easily disambiguate the referent, and (b) preferentially distinguish frequently referenced colors.

Let's start with perceptual structure. We can represent colors in a three-dimensional space that captures the "shape" of color perception, where distance in space reflects perceptual discriminability. The distribution of colors in this three-dimensional space is irregular—it is "bumpy" and has at least

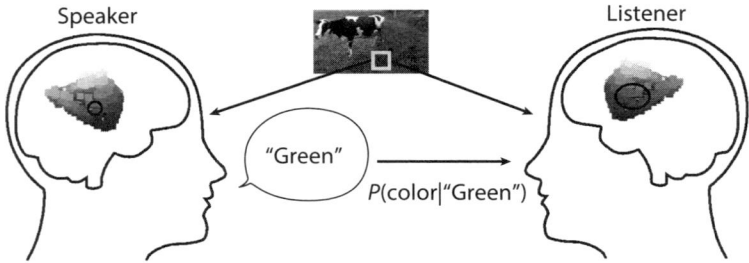

FIGURE 11.1. The color communication problem. A speaker wishes to refer to a particular color using a color term. The listener hears the term and forms (using Bayes' rule) a posterior distribution over colors.

one large protrusion around red/yellow. One implication of this irregular distribution is that colors sitting on the bumps will be more discriminable because they have fewer neighbors. This is conspicuous with respect to Berlin and Kay's typology; if you have only four color terms, then you should include red and yellow, because these colors are relatively easy to disambiguate for listeners. Thus, communicative efficiency places pressure on color categories to delimit perceptually distinct regions of color space.[17]

As we've already seen, informativeness and effort necessarily trade off. Different languages evidently choose this trade-off in different ways, with some languages demanding less effort (fewer color terms) and conveying less information. There are many ways to satisfy this trade-off, but only some of them are maximally efficient in the information-theoretic sense. A language could theoretically choose boundaries that do not maximize information for a given level of effort (we've already seen how Himba and Berinmo demand the same level of effort but choose their boundaries in slightly different ways). Maximally efficient languages can be said to live on an "efficient frontier" that maximizes information for a given level of effort. It turns out that most languages live very close to this frontier, and variation between languages can be captured by a single parameter that governs the informativeness-effort trade-off (Figure 11.2).

Let's now turn to communicative need. This is difficult to estimate for colors, since we have no way of objectively measuring how much a particular color is needed for communication. One proxy is the color distribution of foregrounded objects in natural scenes, assuming that people typically refer to the colors of objects that they see in front of them. Foreground colors tend to be "warm" (yellow/orange/red), whereas background colors tend to be "cool" (blue/green).[18] Warm colors also tend to be more easily identified on

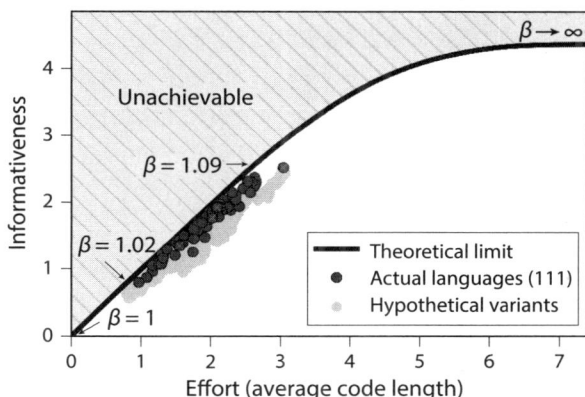

FIGURE 11.2. Color naming systems mostly lie along the efficient frontier. The parameter β_l, estimated for each language, governs the trade-off between informativeness and effort. Adapted from Zaslavsky et al. (2018).

the basis of color labels; in other words, the color categories more finely partition the warm region of color space, and this is true for multiple languages. Putting these observations together, we can infer that natural languages pay a greater coding cost (more bits to represent a finer partition) for colors with high communicative need.

The analysis of communicative need can also shed light on cross-linguistic variation. Although this is hard to study for color (there aren't enough text or speech corpora to estimate word frequencies in many languages), it has been studied for other kinds of semantic categories. A well-known (and somewhat infamous) example is the observation that some Eskimo languages assign different terms to subtypes of snow.[19] While the number of Eskimo terms for snow has been greatly inflated by popular culture, recent analyses have confirmed that there is a kernel of truth to this observation: languages tend to have different terms for snow and ice if they are spoken in colder climates.[20] More generally, many aspects of languages seem to derive from properties of speakers' ecological, social, and technological niche.[21]

I have described perceptual structure and communicative need as two different factors in determining semantic categories. However, these factors are not strictly separate. It is conceivable that perceptual structure is shaped by communicative need. The bumps on the surface of perceptual color space, for example, might not be arbitrary, but instead might emerge from the fact that certain colors are needed more than others.[22]

Word-order puzzles

In this section, I describe two word-order puzzles. First, why are certain orderings of subject/object/verb preferred over others? Second, why are some adjective orderings preferred over others? I will show that principles of efficient communication provide an answer to both puzzles.

Who did what to whom?

In English, if you want to express the idea that a person (the subject) is acting upon something or someone (the object), you will use subject-verb-object (SVO) order, such as "Lisa [S] pushed [V] John [O]." However, another order (subject-object-verb, SOV) is more common across languages. A much smaller number of languages use verb-subject-object (VSO). SOV may have been the historically default word order from which SVO languages evolved.[23] Indeed, if you take speakers of SVO languages and have them develop gestural communication systems, they will tend to produce systems with the SOV order.[24]

Why do languages prefer certain orders over others? In this section, I will describe an efficient communication account of word order developed by the psycholinguist Ted Gibson and his collaborators.[25] If there is any noise in the communication channel (either from the speaker making a mistake or the listener failing to attend), the listener must try to correct the distortions. A language designed for efficient communication will build in some redundancy that allows distortions to be corrected more effectively. Evidence suggests that speakers choose their utterances to protect primarily against deletions of words, compared to insertions and transpositions.[26]

Different word orders vary in their sensitivity to particular distortions, depending on sentence structure. For semantically non-reversible sentences like "Lisa pushed the table," in which only Lisa can occupy the subject role, word order does not help protect against deletions; if "Lisa" is deleted, it is unambiguous that "the table" cannot be the subject, since tables cannot push, regardless of SVO or SOV order. However, for semantically reversible sentences like "Lisa pushed John," in which both Lisa and John can occupy either subject or object roles, SVO will allow the listener to disambiguate whether "John" is subject or object, whereas SOV will not. If I am a speaker of an SVO language, all I have to do is check whether "John" comes before or after "pushed." But if I'm a speaker of an SOV language, the partial sentence "John pushed" is equally compatible with the reconstructions "John Lisa pushed" and "Lisa John pushed."

The puzzle is why SOV is the preferred word order for most languages if it is actually more ambiguous than SVO. This would seem to argue directly against the efficient communication hypothesis. There is another piece of the puzzle, however. Many languages have another means of communicating role assignments: case marking. If you translate "Lisa pushed John" into an SOV language like German, the two nouns would have different forms depending on which role they occupied. This is sufficient to disambiguate roles after deletion. This syntactic redundancy means that additional constraints, such as word order, would require needless effort for the speaker, and hence should be eliminated. Consistent with this hypothesis, languages with stronger case marking show weaker word order constraints.[27] Gibson and his collaborators have also shown experimentally that when speakers communicate gesturally about reversible sentences (where SOV is not necessary to disambiguate after deletion), they prefer SVO order.[28]

Adjective order

Phrases like "small red car" and "heavy wooden box" are preferred to "red small car" and "wooden heavy box" in English. Clearly these phrases convey the same semantic information, so why the order preference?

One clue is that adjective order can be partly predicted by subjectivity (the degree to which different speakers will tend to agree about a property of the world). Adjectives like "small" and "heavy" are subjective because they vary depending on contextual factors (e.g., the distribution of sizes and weights for a set of objects) that might be specific to a speaker, whereas adjectives like "red" and "wooden" tend to be relatively more context-invariant.[29] More subjective adjectives tend to be placed farther away from the noun.[30] So now we have a different puzzle: Why does adjective order depend on subjectivity?

Sentence comprehension is limited by memory: more recent words are remembered better. Thus, if speakers try to design utterances that protect against distortion, they should choose word orders in which loss of older words is minimally destructive to the meaning of the sentence.[31] In adjective-noun phrases, the goal is to place adjectives closer to the noun if those adjectives are more informative about the meaning that the speaker is trying to convey.

To illustrate how this relates to subjectivity, imagine that a moving company arrives at a house to load boxes into their truck. Mover A says to mover B, "Go pick up the heavy wooden box." If mover B only caught the end of the sentence ("wooden box"), she could narrow down the set of candidate boxes by eliminating from consideration all the non-wooden ones. But if mover A had said, "Go pick up the wooden heavy box," then mover B is stuck with figuring out which boxes are heavy. This is more ambiguous, because heaviness is subjective: a given box could be designated heavy or light depending (for example) on the speaker's strength and their beliefs about what's inside the boxes. Thus, putting more objective adjectives closer to the noun is a property of well-designed utterances under the assumption that listeners are memory-limited.[32]

Language learning and evolution

You have probably never heard the sentence "The purple zebra climbed on top of the rocket ship," yet you have no trouble understanding what it means. The fact that our knowledge of language allows us to generalize limitlessly raises several questions:

- **The representation question**: What allows language to encode an infinite number of meanings from a finite vocabulary?
- **The learnability question**: What allows us to learn an infinitely expressive language from finite data?
- **The evolution question**: What is the origin of infinitely expressive linguistic structure?

In addressing these questions, I will focus on the constraints imposed by communicative efficiency. The goal for speakers is to communicate as many meanings as possible with the least effort, and the goal for listeners is to understand as many meanings as possible with the least effort. These twin goals place pressure on languages to be both *compressible* (the rules of language should have a short description length) and *expressive* (the rules of language should be able to generate a wide variety of expressions with distinct meanings). Compression trades off against expressivity: if a language is too compressed, then it becomes ambiguous, with multiple expressions mapping onto the same meaning. (As we saw above, language can tolerate ambiguous word meanings so long as the ambiguity is mostly resolved by context, but this simply means that language as a whole cannot compress both word meaning and context too aggressively.)

In the information-theoretic framework, effort is conceptualized in terms of description length—bits are costly to store and manipulate. But this notion of effort also has implications for learning. As I discuss below, there is a deep connection between compressibility and learnability. Because only learnable languages will be transmitted across generations, compressible languages will tend to proliferate.

Representation

The only way to construct an infinitely expressive system out of a finite number of components is using compositionality (which I first introduced in Chapter 3), whereby the meaning of a complex expression is a function of the meaning of its parts and the structure-building operations that govern their combination. If one can build complex expressions out of simpler expressions, then one can continue iterating these structure-building operations ad infinitum. Technically, to have infinite expressivity, a language needs a stronger form of compositionality, namely *recursion*: structure-building operations that can invoke themselves. For example, the expression "Dan thinks that cats are stupid" could be constructed from the structure-building operation $\texttt{thinks}(X, Y)$ that takes as input X and Y, returning a new expression "X thinks that Y." This operation can invoke itself in the input, as in $\texttt{thinks}(X, \texttt{thinks}(A, Y))$, which returns "X thinks that A thinks that Y" ("Dan thinks that Bill thinks that cats are stupid"), and this process can continue ("Dan thinks that Leslie thinks that Bill thinks that cats are stupid"). It has been argued that recursion is the critical ingredient of language that distinguishes it from other animal communication systems.[33] However, compositionality without recursion can still potentially produce a very large (albeit finite) set of meanings from a small number of components and operations.

We can contrast compositional languages with *degenerate* and *holistic* languages.[34] In a degenerate language, all meanings are mapped to the same utterance. Such a language achieves maximum compression and is thus easily learnable, but at the cost of extreme ambiguity (minimum expressivity), making it communicatively useless. The character Groot in the comic book series *Guardians of the Galaxy* appears to speak a degenerate language (only uttering "I am Groot"), though some of the other characters seem able to decode finer shades of meaning. In a holistic language, the mapping from meaning to utterance has no internal structure; two different utterances, arbitrarily close in meaning, may be arbitrarily different in their surface forms. Such a language minimizes ambiguity (maximizes expressivity), but does not permit any compression. The most efficient encoding of a holistic language is simply a dictionary, which does not support any generalization to novel utterances, thus making it unlearnable. Compositionality strikes a balance between expressivity and compressibility.

Learnability

As was discussed in Chapter 2, learning without an inductive bias is not generally possible, because finite data cannot disambiguate all possible hypotheses. The need for inductive bias motivated the idea of *universal grammar*, a restriction on the hypothesis space for syntax that would allow it to be learnable.[35] Various theoretical results, starting from different assumptions but arriving at more or less the same conclusion, support the conclusion that language is unlearnable without an inductive bias.[36]

The inductive bias needn't be a hard restriction on the hypothesis space, as in the original version of the universal grammar hypothesis. It can also be "soft" in the form of a prior distribution that favors some languages over others.[37] In the probabilistic setting, there is an inverse relationship between the amount of data needed to learn the correct language and the prior probability of that language.[38] Recall from the last chapter that high probability hypotheses will generally have a shorter description length—they are more compressible under an entropy encoding scheme. Thus, compressibility translates into learnability, indicating a synergy between approximation and inductive biases.

Evolution

Language is a cultural artifact, transmitted across generations through verbal communication. What kinds of languages will this transmission process tend to produce? We can provide an analytical answer to this question if we make the simplifying assumption that transmission is "vertically" organized into a chain, where a single listener receives linguistic data from a single speaker, makes an inference about the language generating that data, and then samples new data from the inferred language for the next speaker in the chain (Figure 11.3). This chain will converge to a language that is high probability under the prior distribution over languages.[39] In other words, inductive bias not only determines the learnability of a language, but also shapes the language itself.

The analysis of transmission chains suggests that languages should become increasingly degenerate over time if the inductive bias favors simpler (more compressible) languages. However, this ignores the fact that real speakers do not simply sample random utterances from the inferred language, but rather use these utterances to communicate information about the world or about their mental states. To do this, speakers need an expressive language. Thus, if speakers are informative, then transmission will produce languages that negotiate a compromise between expressivity and compressibility—precisely the class of compositional languages.[40]

This prediction has been borne out not only theoretically but also in laboratory analogs of cultural transmission. The cognitive scientist Simon Kirby and his collaborators have developed an experimental paradigm in which people are asked to communicate information about novel pictures using strings of letters.[41] The first "generation" of speakers is exposed to a set of pictures

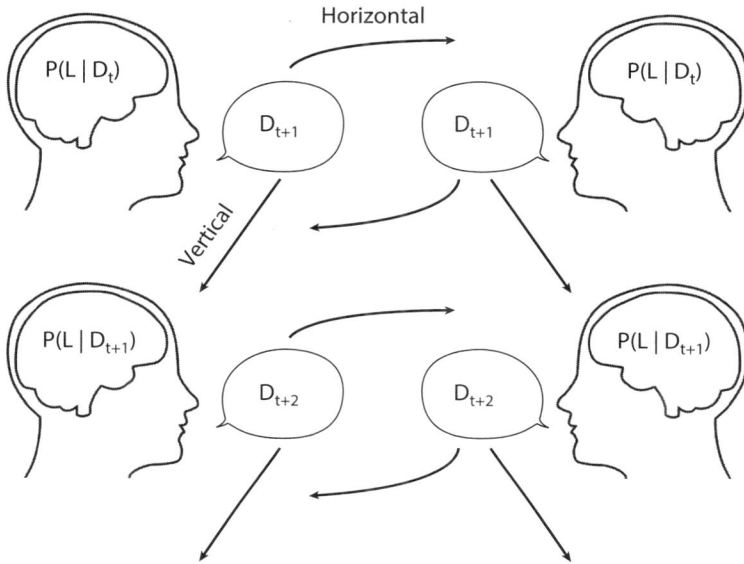

FIGURE 11.3. Vertical and horizontal cultural transmission. At iteration t, an individual receives linguistic data D_t, infers the generating language L, and generates data D_{t+1} by sampling from the inferred language. The generated data is passed on to the next generation (vertical transmission) and/or circulates between speakers of the current generation (horizontal transmission).

along with corresponding letter strings, and then partners play a communication game in which the pair is rewarded if the listener can successfully identify a picture on the basis of the speaker's chosen string sequence. One of these string sequences is then used to produce data for the next generation. Consistent with the theory outlined above, Kirby has shown that this transmission process tends to produce increasingly degenerate languages over successive generations. However, when transmission is organized horizontally, whereby the same people participate across multiple generations, the resulting languages tend to be compositional, with substrings that are reused to convey systematic variations in meaning (e.g., texture and shape).[42] This finding was anticipated by simulations of cultural transmission, which showed that horizontal transmission places weaker pressure on learnability (and hence relatively stronger pressure on expressivity), due to the fact that learners will be less reliant on their inductive bias over successive generations by virtue of repeated exposure.

Summary

This chapter covered five broad linguistic issues: ambiguity, meaning, word order, learning, and evolution. For each one, I showed how efficient communication supplies a coherent explanatory framework. The normative prescription is that language should communicate as much as possible with as little effort as possible, and natural languages seem to largely satisfy this prescription. One conclusion we might draw is that the designers of artificial languages were on the wrong track when they sought to "improve" human language. What they saw as flaws, such as ambiguity and arbitrariness, may in fact have been useful features of communication given the cognitive constraints of speakers and listeners.

12

The uses of randomness

Chance itself pours in at every avenue of sense.

—C. S. PEIRCE

We usually think of well-designed systems as being completely deterministic and predictable. When I flip the light switch, I expect the light to turn on. If it turns on and off unpredictably, I know there's a problem. But this intuition does not always work. Sometimes it's not possible to build a system that is both reliable and deterministic.

To understand why, we need to separate the concepts of reliability (consistently performing a particular task) and determinism (consistently doing the same thing). I can reliably perform a task without being deterministic. For example, suppose you pay me to mow your lawn once a week. Every week you come home from work and see that your lawn is perfectly mowed—I've performed my task reliably. But if you watched me do it, you would find that (to avoid boredom) I do it differently each time. Some weeks I mow in horizontal stripes, sometimes in vertical stripes, sometimes in concentric circles, and so on. There's no way for you to predict one week how I'm going to do it the next week—I'm non-deterministic.

I've shown that you can be both reliable and non-deterministic, but what's the point (aside from maybe avoiding boredom)? The reason non-determinism is important from a computational point of view is that sometimes the *only* way to build a reliable system is to be non-deterministic. In this chapter, I will survey four examples illustrating the different ways that the brain harnesses the power of randomness.

Sampling beliefs

In 1946, the physicist Stanislaw Ulam was convalescing from an illness, trying to keep himself entertained by playing solitaire. His fertile mathematical mind, unable to tolerate bed rest, spun itself the following question: What are the chances that a game of solitaire will come out successfully? Being a physicist, he set about trying to calculate this probability using combinatorics, but that proved too difficult. Then he had a flash of insight:

> I wondered whether a more practical method than "abstract thinking" might not be to lay it out say one hundred times and simply observe and count the number of successful plays. This was already possible to envisage with the beginning of the new era of fast computers, and I immediately thought of problems of neutron diffusion and other questions of mathematical physics, and more generally how to change processes described by certain differential equations into an equivalent form interpretable as a succession of random operations.[1]

At the time, Ulam was working on nuclear weapons research at the Los Alamos Scientific Laboratory. In their investigations of radiation shielding, Ulam and his colleagues were trying to figure out how far a neutron would travel through some material until it collided with an atomic nucleus, releasing some amount of energy. This problem resisted analytical calculation, but Ulam realized that by simulating neuron diffusion many times on a computer, he could get a reasonably accurate estimate of the desired quantity. Mathematical purists might not be satisfied, but Ulam's technique got the job done. The writer George Dyson described the physicists' view of the technique as "a form of emergency first aid, in answer to the question: What to do until the mathematician arrives?"[2] They code-named the project "Monte Carlo" after the casino in France where Ulam's uncle liked to gamble (the same casino that inspired the gambler's fallacy in Chapter 7).

To gain some intuition for how and why Monte Carlo sampling works, imagine a jar filled with colored marbles. You don't know how many marbles of each color are in the jar, and let's assume that there are too many to count. If you reach in and grab a random sample of the marbles, then you can easily count the proportion of each color in the handful—this is the Monte Carlo estimate of the color distribution. One can show mathematically that if you repeatedly draw handfuls of marbles and count the color proportions, then the average of these proportions will converge to the true proportions in the population of marbles. This is the same logic underlying the use of representative samples when conducting polls: assuming we're randomly sampling representative subsets of the population, then we can be confident that these samples will on average reflect the true population distribution. Moreover, you can apply this logic to more than just estimating distributions. Any function of the distribution, such as the median or most frequent color, can be estimated from a random sample.

The Monte Carlo technique turned out to have revolutionary implications far beyond nuclear physics. Statisticians and computer scientists now routinely use it to approximate otherwise intractable inference problems, and have developed many sophisticated variations on Ulam's original technique.[3] The ubiquity of Monte Carlo has naturally led some cognitive scientists to ponder whether it might also be used by the brain. In fact, thinking about the brain as implementing some form of Monte Carlo technique can illuminate many otherwise puzzling phenomena.

Noisy neurons

In Chapter 10, I discussed noise in neural coding as an unavoidable nuisance given the brain's limited resources. Neurons are noisy devices,[4] and the channel design problem is a way to cope with this noise. For a fixed input, neural firing approximately follows a Poisson distribution, emitting an action potential with some constant probability that doesn't depend on the time since the last action potential.[5] The Poisson distribution is the most random (highest entropy) distribution for a given firing rate. This means that neurons are not just noisy—they're as noisy as a channel can be while still communicating information at a fixed rate. This makes sense from an efficient coding perspective, since reducing variability (e.g., by opening more ion channels, or increasing neurotransmitter volume) is metabolically costly.[6]

Note that "noise" here doesn't have to mean *truly* random, in the sense that the underlying physical process is non-deterministic. True randomness might only appear at the level of quantum mechanics. What I mean by "noise" is that some variable cannot be predicted by an observer given the information available to the observer.[7] In the case of neurons, it is known that action potentials are actually highly predictable if one measures a neuron's membrane potential.[8] But because a downstream neuron (the "observer") has no access to the upstream neuron's membrane potential, the action potentials are effectively random from the perspective of the downstream neuron.

The task facing the downstream neuron is to decode the hidden variable (the message) conveyed by the upstream neuron's firing. The ideal decoder would follow Bayes' rule, combining the prior probability of each hypothetical message with its likelihood given the noisy code. It is possible to construct neurons that do this, where the firing of the downstream neuron signals its posterior belief about the hidden variable encoded by its inputs.[9] However, we have ignored a key fact about the computational problem: the prior probability may depend on information conveyed by other noisy neurons, and multiple neurons may be reciprocally connected in a network. This network will be in constant flux, even when the sensory data are held fixed. So how can the neurons stably report posterior beliefs?

Consider, for example, the problem of edge detection. Images consist of partially overlapping object surfaces, and the problem is to detect the edges that delimit surface boundaries. If we have N image regions, each containing at most one edge, and K possible edge orientations, then we have K^N possible configurations of edges. This can be an astronomical number even for a modest number of regions and orientations; with only 15 regions and 10 orientations, there are a trillion configurations. Doing exact Bayesian inference in this configuration space is clearly impossible. If every edge was independent, then this problem would be a lot simpler, since we could just assign K neurons to each region. But of course edges aren't independent; they tend to form smooth contours due to the fact that they arise from physically extended objects. In other words, the prior probability of an edge in a particular region will depend on the edges of its neighbors. This is important, because edges are often ambiguous. Two surfaces may have similar color and texture (recall the Kanizsa triangle in Figure 2.4, where we see a vivid foreground triangle despite the absence of visible edges). Our visual system resolves this ambiguity by exploiting the smoothness and locality of edges in natural images.[10] The primary visual cortex contains a population of neurons ("simple cells") that function as simple edge detectors, firing when they believe the visual input contains an oriented edge in a particular location. Because a single simple cell, tuned to a small region of space, cannot resolve edge ambiguity on its own, it receives excitation from other cells tuned to nearby regions, and reciprocates this excitation, such that the response of the cell is enhanced when its neighbors detect an edge with similar orientation, thus forming a locally smooth contour.[11]

Edge detection is a chicken-and-egg problem: if I'm a neuron trying to detect an edge in a particular region, I need to know the orientation of edges detected by neurons tuned to neighboring regions, but my neighbors are just as clueless as I am, since they're dependent on information coming from their neighbors (including me). A clever way to crack this problem is by starting from the observation that if a neuron hypothetically had access to the ground truth for its neighbors, then it could compute its local posterior exactly. If each neuron draws a sample from its local posterior distribution, firing with probability proportional to its belief conditional on the samples generated by its neighbors, then the network will instantiate a dynamical system that evolves until it reaches a state of equilibrium. It turns out that at equilibrium, we can interpret the population activity as a sample from the joint posterior over configurations.[12] In other words, instead of trying to exhaustively enumerate the K^N possible configurations (intractable) or treat all edges as independent (too severe an approximation), the network can stochastically sample from the set of configurations. Sampling works in practice because even though the number of configurations is vast, the number of high probability configurations is much smaller, and stochastic sampling preferentially explores this high probability set. The neural population is approximating the inference problem in much the same way that Ulam alighted upon when thinking about solitaire and neutron diffusion.

If this account is correct, then noise in the brain serves a functional purpose, driving sample-based approximate inference. A number of implications flow from this basic insight. First, consider "spontaneous" activity that occurs in neural populations when no stimulus is present (of course, there's no such thing as truly spontaneous activity, since the brain is always getting some sensory input as long as we're awake, so this should be interpreted narrowly to mean "spontaneous with respect to the experimental procedure"). Over the course of visual experience, the distribution of spontaneous activity in the visual cortex comes to increasingly resemble the distribution of stimulus-evoked activity. From the stochastic sampling perspective, we can interpret this spontaneous activity as reflecting the prior distribution over scene variables (e.g., edges). The shift towards the evoked distribution happens because the posterior at one point in time becomes the prior at the next point in time.[13] A second implication, supported by experimental data,[14] is that if you repeatedly present stimuli sampled from the distribution of natural images, the average evoked distribution will look like the spontaneous distribution. This follows from the fact that the average posterior is just the prior:

$$\sum_d P(h|d)P(d) = P(h), \tag{12.1}$$

where h denotes the hypothesis (e.g., edge configuration) and d denotes the data (e.g., images). A third implication is that variability of neural firing should reflect uncertainty. When stimuli are presented, uncertainty (and hence variability) should decrease. This is what is seen in experimental studies: stimulus onset quenches neural variability.[15]

Perceptual signatures of Monte Carlo

If the brain approximates Bayesian inference by sampling, we should be able to see signatures of this sampling process in perception. Sampling seems particularly well-suited to explain multi-stable stimuli such as those shown in Figure 12.1. Each stimulus can be interpreted in two ways (there are some stimuli, not shown here, that can be interpreted in more than two ways). The Necker cube can be interpreted in two different three-dimensional

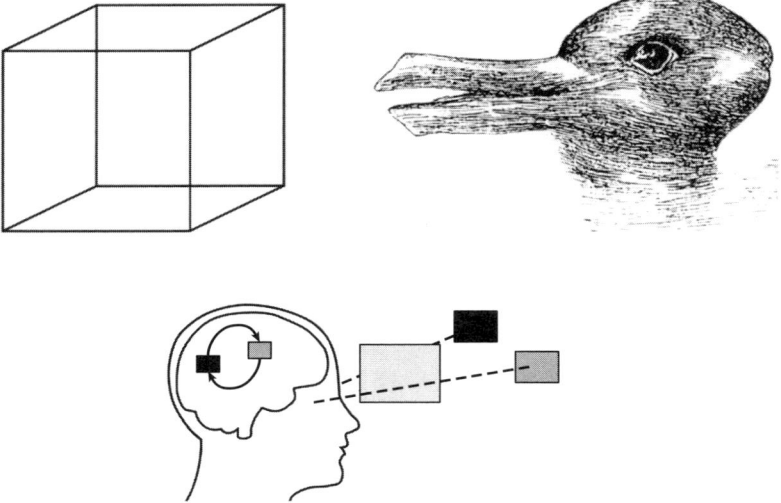

FIGURE 12.1. (Top left) The Necker cube. (Top right) The duck-rabbit illusion. (Bottom) Binocular rivalry.

configurations, and the duck-rabbit illusion can be interpreted as a duck or a rabbit. Another extensively studied multi-stable phenomenon, binocular rivalry, uses stereoscopic presentation, where each eye sees a different stimulus. When you look at these stimuli, your conscious perception oscillates back and forth between the two interpretations (or the two different images in the case of binocular rivalry).

Why does perception oscillate? One thing that all multi-stable stimuli have in common is that the different interpretations are all roughly equally consistent with the visual input. This means that the posterior distribution over interpretations has two peaks. The neural population representing the different interpretations will therefore randomly meander between the interpretations as it generates samples from the posterior.[16]

This theory raises an interesting question: If the populations are meandering between the interpretations, why don't we experience intermediate perceptual states in which we see mixtures of the stimuli? The answer is that if the posterior is peaky enough (with very little probability on the intermediate interpretations), then these will be rarely sampled. However, under some conditions (e.g., under low contrast or when the two stimuli in a binocular rivalry experiment are similar[17]), these intermediate interpretations have higher probability, and people do report seeing them.

Another question is why the stimuli seem to switch as a whole rather than in a piecemeal fashion. The answer is that it depends on the strength of probabilistic coupling between different parts of the image. For many objects, it would be weird if one part of the object suddenly changed to something completely different. You don't typically come across animals that are part duck and part rabbit. And you never come across cubes with vertices that simultaneously occupy two different locations in three-dimensional space. But under some conditions (e.g., if the stimuli are sufficiently large[18]), the coupling is weak and you get "piecemeal" switches.

The coolest example of piecemeal rivalry is the phenomenon of *traveling waves*.[19] When two different circular gratings are presented to the eyes, at any given moment one of the gratings will be dominant (accessible to conscious awareness) and one will be suppressed, as in standard binocular rivalry experiments. Transiently increasing the contrast at one location of the suppressed grating effectively lights a fuse that causes the suppressed grating to reveal itself dynamically, starting with the high-contrast location and traveling along the grating. By measuring conscious perception at different probe locations along the grating, it is possible to reconstruct the speed at which these waves travel, and to show that the wave is slowed down when there is a gap in the grating. Remarkably, these waves literally travel across the surface of visual cortex (whose neurons are topographically mapped as a function of visual location).[20] These findings demonstrate that randomness in visual perception is constrained by stimulus structure.

Posterior probability matching

Many of the studies described in this book report behavior consistent with Bayesian inference. But there's a catch: this consistency is typically assessed at the *aggregate* level, after averaging behavior across many individuals. This is done because an individual's data are often very noisy, making it difficult to identify regularities. Nonetheless, this method disallows assessment of the hypothesis that individuals are Bayesian. In fact, close inspection of individual behavior reveals that they are not doing exact Bayesian inference.[21] The consistency at the aggregate level is an averaging artifact.

A few examples will illustrate this point. The psychologist Robert Nosfosky and his collaborators ran an experiment in which people learned categories defined by binary features (rocket ships that varied in the shape of different component parts, where each part can only take one of two shapes), and then were asked in a transfer test which category new rocket ships (which they had never seen before) belonged to.[22] In reanalyzing these data, Noah Goodman and his collaborators observed that categorization judgments on the transfer test very closely matched the predictions of a Bayesian category learning model (Figure 12.2).[23] However, notice that what this graph is really showing is that the *proportion* of people making a particular category

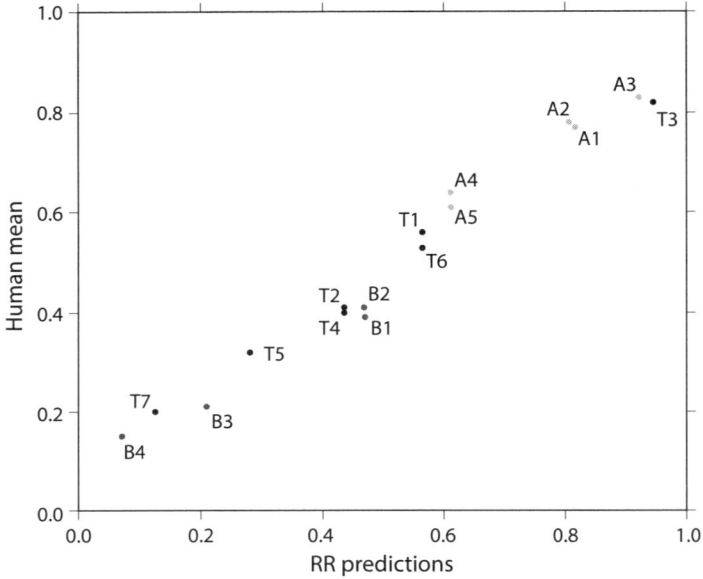

FIGURE 12.2. Average categorization performance almost perfectly matches categorization probabilities from a Bayesian model (the "rational rules," or RR model). Reproduced from Goodman et al. (2008).

judgment matched the posterior probability from the Bayesian model. This does not tell us anything about what the individuals are doing. When the data are disaggregated, one sees that there is a huge amount of variability across people (a point already observed by Nosofsky). In particular, people seem to be reporting categorization judgments with probability proportional to the posterior, a phenomenon known as *posterior probability matching*.[24] If people were instead doing exact Bayesian inference, their judgments would be completely deterministic (modulo some response noise).

The second example is taken from a study by Tom Griffiths and Josh Tenenbaum.[25] In this study, people were asked to make predictions about intuitive quantities, such as "What is the total baking time of a cake given that it has been baking for 45 minutes?" Human judgments closely matched the posterior median on average. However, this did not hold true at the individual level, where judgments appeared to be more consistent with the assumption that they were random samples from the posterior.[26]

Children also exhibit posterior probability matching. In an experiment carried out by psychologist Stephanie Denison and her colleagues, children (four and five years old) were told about the causal properties of blocks distinguished by their colors. Some blocks turned a novel sound machine on, and some didn't. The children then learned about the proportion of different colors in a bucket. Finally, the children were told that one of the blocks from the bucket accidentally turned the machine on, and were asked to identify the color of the block they thought was responsible. By varying the proportion of colors in the bucket, the experimenters were able to parametrically change the posterior distribution for the errant block in the test phase. They found that the number of children reporting a color was proportional to that color's posterior probability.[27]

Posterior probability matching is broadly consistent with the idea that people approximate Bayes' rule by sampling. Note that the sampling hypothesis is only consistent with posterior probability matching when the number of samples is small; as the number of samples increases, the Monte Carlo approximation should come to increasingly resemble the exact posterior, and then judgments based on this large-sample approximation will be indistinguishable from exact inference. So the question becomes: Why do people only take a small number of samples?

How many samples?

Sampling is costly in several ways. First, it may demand cognitive resources to produce samples. When people are placed under cognitive load (for example, by asking them to simultaneously perform a secondary task), which putatively reduces the availability of cognitive resources, they appear to generate fewer samples.[28] Second, when sampling unfolds over time, there is an opportunity cost to generating more samples; time spent sampling could alternatively be spent acting. A basketball player can improve her chances of scoring by spending more time sampling, but if she spends too much time, she might miss the buzzer. The optimal number of samples will depend on the relative costs and benefits of sampling. In many cases, good decisions can be made on the basis of surprisingly few samples.[29] This suggests that we should see behavior consistent with drawing a small number of samples.

One consequence of small samples is that the distribution may not reach equilibrium (i.e., samples will not be drawn from the posterior distribution). If this is the case, then the samples will be biased by their history. This will produce "anchoring" effects, where seeding people with an initial guess, even when irrelevant, can systematically tip their judgments in the direction of the guess. In their classic studies of anchoring, Amos Tversky and Daniel Kahneman asked people to estimate quantities like the percentage of African countries in the United Nations, and produced manifestly irrelevant anchors by spinning a wheel of fortune. People first had to judge whether the quantity was higher or lower than the anchor, and then give a direct estimate of the quantity. In principle, they could have just ignored the relative judgment when making the absolute judgment, but instead the anchor biased their responses, such that the estimates were lower when the anchor was lower. Anchoring is not just a quirk of this rather odd experimental setup. It has, for example, impacts on bargaining (final settlements are biased towards initial offers[30]) and criminal sentencing (judges are biased by sentencing demands, even when coming from an irrelevant source[31]).

The cognitive scientist Falk Lieder and his collaborators have explored anchoring effects within a sampling-based framework, where the critical new element is that the number of samples is adaptively determined by cost-benefit analysis.[32] I'll briefly summarize a few of the ways this analysis illuminates the determinants of anchoring effects. First, placing people under cognitive load should increase the cost of sampling, and hence amplify the anchoring effect, consistent with experimental evidence.[33] Second, the model predicts that increasing incentives should motivate people to sample more, thus reducing the anchoring effect. This incentive effect is stronger when people have greater knowledge of a domain, and hence can benefit more from additional samples.[34]

Anchoring effects also show up in probability judgments. Suppose you visit a sketchy buffet, and you wonder to yourself, "What's the probability that I'll get sick after eating here?" This question is mathematically equivalent to the question, "What's the probability that I'll get sick after eating the salad, or after eating the noodles, or after eating the soup . . . ," and so on, enumerating each possible individual way of getting sick. Despite their mathematical equivalence, people tend to underestimate more in their answers to the first question compared to the second question. Intuitively, if I don't enumerate the possibilities to you, then you have to self-generate them (e.g., by sampling), and this enumeration may be incomplete due to the cost of sampling. Studies in my lab have found greater underestimation when people are incentivized to respond quickly, consistent with Lieder's cost-benefit analysis.[35]

To summarize: randomness is not random. People appear to be quite sophisticated in their randomness, calibrating the number of samples they need to do well on a task relative to the costs of generating those samples.

The exploration-exploitation dilemma

In the last section, I assumed that people have some control over how they sample hypotheses given some fixed data. But recall from Chapter 5 that people also have control over how they sample data. In that chapter, I discussed models of data selection based on the principle of information maximization, and how these models can predict some counter-intuitive findings (such as the positive test strategy). The underlying assumption is that people are trying to learn as much as possible about the world. That can't be the whole story, of course. People can't devote all their time and energy to gathering data; they also have to *do* things! The challenge is to gather enough information to do things well (i.e., earn reward), without gathering too much information at the expense of reward.

To give a concrete example, let's suppose you move to a new city and you want to identify the best restaurant.[36] For simplicity, let's suppose that you have no preferences for diversity; you'd happily eat at the same restaurant every day if you believed it was the best. After eating at one restaurant, you can't say much about whether it's the best, so you need to try at least one more. Now suppose you like the second restaurant better. Is it the best? There are two issues here. One is that the quality of a given restaurant on any given day may be variable (e.g., different kitchen staff, availability of ingredients, etc.), so you can't be completely confident that the second restaurant is better than the first restaurant *on average*. To figure that out, you would have to sample both of them more times. The second issue is that there could be another restaurant, as yet untried, that is better than both of the restaurants you've already tried. To figure that out, you'd have to explore more restaurants. But in doing so, you run the risk of exploring a lot of duds and paying the opportunity cost of not eating at the restaurants you already knew something about.

This example illustrates the *exploration-exploitation dilemma*: you have to choose between exploiting good options you know or exploring possibly better (but also possibly worse) options you don't know. The restaurant case is a special case of this problem known as the *multi-armed bandit*, which consists of a set of options that deliver rewards from some distribution. The decision problem is to choose a sequence of options (arms) that maximizes reward over some (possibly infinite) time horizon. What makes this difficult is that the optimal choice depends not only on the immediate reward from that choice, but also on the information that it provides. Determining the optimal policy is in general computationally intractable. The mathematician Peter Whittle quipped that during World War II, the problem "so sapped the energies and minds of Allied analysts that the suggestion was made that the problem be dropped over Germany as the ultimate instrument of intellectual sabotage."[37]

Given this intractability, much work in computer science has gone into devising approximate algorithms for solving the exploration-exploitation dilemma in multi-armed bandits. The simplest form of approximation is to introduce some randomness into the choice process. For example, the ϵ-greedy strategy keeps track of the average reward for each option, choosing the

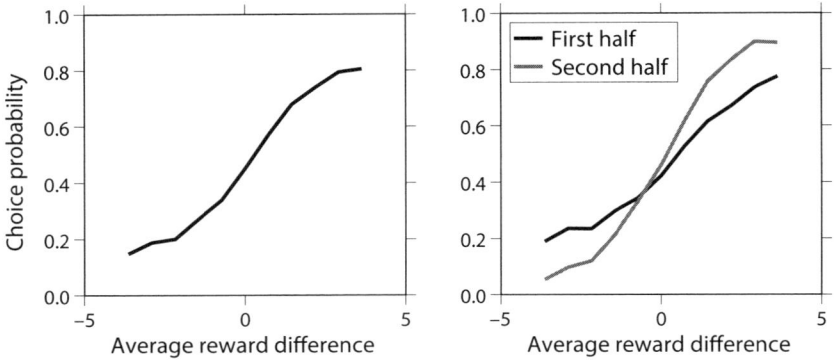

FIGURE 12.3. (Left) Probability of choosing option A over option B as a function of the difference in average reward between A and B. (Right) Same plot, but plotted separately for choices in the first and second halves of each learning block. From Gershman (2018). Copyright © 2018 by Elsevier. Reprinted by permission.

option with the highest average reward with probability $1 - \epsilon$, and choosing a random option with probability ϵ. This strategy explicitly controls the rate of exploitation. Provided that ϵ is greater than 0, following this strategy will allow you to eventually identify the option with the highest reward. However, this strategy is very wasteful, because it doesn't discriminate in a graceful way between options with different average reward. If you tried three restaurants—one great, one good, and one terrible—then ϵ-greedy forces you to choose the good and the terrible options at the same rate whenever you elect to explore.

A slightly more sophisticated strategy, known as *softmax exploration*, is to choose an option with probability proportional to its average reward. Since outcomes could potentially be negative (losses), but probabilities must be positive, we can take the exponential of the average reward to give the probability of choosing option C: $P(C) \propto \exp[\beta V(C)]$, where $V(C)$ is the option's average reward. The parameter β controls the degree of exploration. When $\beta = 0$, choices are random, and as β gets larger, you tend to increasingly choose the option with the highest average reward. The softmax strategy is *value-directed*, in the sense that exploration is targeted at options with higher values, so that you don't forego too much reward when you elect to explore.

When there are only two options (A and B), we can visualize a person's exploration strategy by plotting the probability of choosing A over B as a function of the difference in average reward between the two options (Figure 12.3, left). Consistent with softmax exploration, the choice probability function is S-shaped, flattening out for extreme average reward differences. The exploration parameter β controls the slope of this function, with steeper slopes for larger values of β. Roughly speaking, β tells you how big of a change in choice probability you can expect by changing the average reward difference (moving along the horizontal axis some unit of distance).

There is, however, more structure to these data, which we can reveal by splitting the plot based on the first and second halves of each learning block (people in this experiment played 20 blocks, each consisting of 10 choice trials). Now we see that the slope of the function is steeper in the second half of each block, as though β had increased. This already tells us that people can't be doing the simple softmax strategy with a fixed level of exploration. One possibility is that people use a heuristic where they increase β over the course of each block. This is intuitive, since as you become more confident about how rewarding each option is, you should shift from exploration to exploitation. Subsequent experiments showed that this can't be the whole story: people explore more not only when they have less experience with a set of options, but also when the rewards are more variable.[38]

To understand these findings, we need to go back all the way to the dawn of work on multi-armed bandits. In 1933, William Thompson published the first paper describing the problem, along with a heuristic solution that has since come to be known as *Thompson sampling*. The idea is simple and elegant. Each option has some long-run average reward (its value). Given a set of rewards generated by an option, we can use Bayes' rule to compute the posterior distribution over value for that option. Thompson proposed that we should choose by drawing a random sample from this posterior, giving us one sampled value for each option, and then choosing the option with the highest sampled value. Thus, you'll tend to choose options with higher average reward, but you'll also explore other options in a graded way, just like in softmax. The critical difference is that when you have more posterior uncertainty, you'll explore more. This can explain why people seem to be more exploratory early during learning or when rewards are more variable. It can also explain why the variability of neural activity encoding value estimates is correlated with uncertainty.[39]

Although Thompson sampling is a heuristic, it turns out to have quite strong theoretical guarantees.[40] In particular, one thing you'd like to know if you use a particular exploration strategy is how much *regret* you'll have on average (the difference between how much you could have earned if you always made the optimal choice and how much you actually earned). In general, regret will grow as you make more choices, since there are more opportunities to make a suboptimal choice. The question is how quickly your regret grows over time. If regret is linear in time, this means that you're missing (on average) a fixed amount of reward on each choice. This is a bad regime to be in, because you're not really reducing the gap over time. Softmax exploration has linear regret; even after you've learned the correct values, you're still choosing stochastically and hence giving up a fixed average amount of reward on each choice. Thompson sampling, in contrast, has *sublinear* regret, which means that you're giving up less and less as you gain more experience and become less exploratory.[41] Thus, we see another rationalization of posterior probability matching, this time from a decision making perspective.

Strategic randomness

So far, I've been dealing with the conditions under which an individual, acting alone, will rationally exhibit randomness (in beliefs or choices). A new dimension emerges when we consider interactions between individuals. The standard mathematical tool kit for understanding such interactions is game theory. In this section, I will informally describe how randomness can be an optimal game-theoretic strategy.

Imagine you're a goalie in a soccer penalty shoot-out. Which direction do you jump? If you know that the kicker has a tendency to shoot towards the left, then you should jump towards the left. But a more sophisticated kicker would know that their predictability can be exploited by the goalkeeper. Likewise, a sophisticated goalkeeper would know that their predictability can be exploited by the kicker. This creates a strategic equilibrium, where no player can improve their chances by adopting a deterministic strategy (formally, this is known as a *mixed strategy Nash equilibrium*). Thus, you should be as random as possible.

Are people able to produce completely random behavior? Despite the evidence reviewed earlier that there's quite a bit of randomness in the brain, some studies have suggested that it's actually quite hard for people to be unpredictable.[42] For example, people seem to believe that long runs of the same move indicate non-randomness, so they tend not to produce such runs, consequently underestimating their occurrence in true random sequences. However, if you put people into a competitive setting (like the penalty shoot-out) with real monetary payoffs, then they are indeed capable of producing nearly random behavior.[43]

This conclusion extends to real penalty shoot-outs. Both goalies and kickers act near randomly, consistent with the fact that their probability of winning is the same no matter which direction they jump or kick.[44] Similar results have been obtained for serve direction in tennis matches.[45] When stakes are high and one can't gain an advantage by being deterministic, professionals are capable of being strategically random.

Beyond sports, examples of strategic randomness have been observed across the animal kingdom.[46] For example, laboratory technicians had long observed that accidentally jangling keys could induce "audiogenic seizures" in laboratory rats. However, if the rats were provided with hiding places, they would run and hide rather than go into a seizure.[47] This suggests that such seizures may be a randomizing strategy for escape when a deterministic strategy is not available. Similar forms of unpredictability in defensive behaviors have been observed in other species. Moths tumble and loop unpredictably when they detect bat ultrasound.[48] Octopi, cuttlefish, and sea pansies rapidly and randomly change the color patterns on their skin to foil the visual search behavior of their predators.[49] These defensive behaviors may even be proactive. Insects have been observed to exaggerate the randomness of their flight paths even when not under predation,[50] possibly as a kind of "insurance" policy.[51]

These kinds of randomized defensive behaviors remain poorly studied in humans. It is tempting to speculate that a related logic underlies the "madman strategy" executed by Richard Nixon during the Vietnam War. As Nixon put it in 1968, "I want the North Vietnamese to believe I've reached the point where I might do *anything* to stop the war." This strategy was echoed by Donald Trump in 2016, when discussing foreign policy during his presidential campaign: "We have to be unpredictable. We have to be unpredictable, starting now." The basic logic of the madman strategy is to convince your adversary that you are unpredictable, and leverage this perceived unpredictability as a deterrent. Thus, unlike the animal examples given above, the strategy here is not only to escape, but to deter attack.[52]

Summary

Reliable systems can, paradoxically, be built out of unreliable processes. For problems in which exact inference or optimization are intractable, randomness can be an asset, providing a way to efficiently search through the hypothesis space. Random action can also be functional in strategic interactions, where it enables agents to achieve an equilibrium in which no one can do better by adopting a deterministic policy. These different rationalizations may explain why randomness seems to be ubiquitous in the brain.[53]

13

Conclusion

WHAT MAKES US SMART

Out of the crooked timber of humanity, nothing entirely straight can be made.
—IMMANUEL KANT

The brain is evolution's solution to the twin problems of *limited data* and
limited computation. Within our lifetimes, we can only spend limited time
gathering information about the world, and we can only devote our precious
metabolic resources to representing, remembering, and thinking about a limited number of things. To solve these problems, the brain is, and must be,
biased: it directs its activities towards data and computations that it antici-
pates will be useful. For this reason, we are good at some things and bad at
other things. By unraveling the origins of these biases, we can understand
the design principles of the brain. In the closing chapter, I will discuss a few
broader implications of this perspective.

Implications for theories of cognition

The project of rationalizing human cognition undoubtedly strikes some readers as stubbornly recidivist. Didn't modern psychology discredit rationality
as a description of human behavior? Why am I trying to revive a corpse? I
certainly see the appeal of a purely descriptive psychology. If we can accurately describe the mechanisms governing thought and behavior, we might
consider our job done. Principles of rationality, optimality, design, and so
on are just so much Enlightenment baggage that impedes scientific progress.
If one accepts this descriptive view of psychology, the practical question is
whether we can hope to attain accurate descriptions without appealing to
more fundamental principles. The space of possible psychological theories

is infinite. How can we even begin exploring this space unless we have some guidance?

Take, for example, the study of decision making discussed in Chapter 10. Through the painstaking accumulation of experimental data, psychologists and economists were able to piece together a quantitative theory of how decisions are made. This work was groundbreaking and impactful, reshaping economics, public policy, and finance. But despite the success of this theory, it was unable to explain a variety of basic facts about decision making, particularly those that required the form of the value function to change under different circumstances. If we therefore must consider all possible context-dependent value functions, the search would be endless. But fortunately (as I discussed in that chapter) there are more fundamental principles that strongly constrain the form of the value function, accurately predicting how it will change under different circumstances. The point is that even someone who doesn't care about normative foundations for psychology should care about normative theories. They provide much-needed discipline to otherwise intractable modeling problems.[1]

Implications for artificial intelligence

I started this book with a critique of current artificial intelligence systems. Looking back on the ideas presented in this book, what lessons can we draw about the design of systems with human-like intelligence?

Machines, just like brains, must reckon with limits on data and computation. All machines therefore have inductive and approximation biases. The key question for our purposes is how well machine biases line up with human biases. In a number of ways, modern artificial intelligence systems have very different biases from people.[2] For example, when playing video games, people are robust to superficial changes in color, and even to more radical changes like blurring all the objects.[3] At the same time, scrambling all the pixels would make it very difficult for people to learn a game,[4] but some artificial systems would be able to cope with scrambling relatively well. In Chapter 3, I laid out a set of inductive biases that plausibly provides the "startup software" for human intelligence: causality (both physical and mental), compositionality, and object understanding. Recent work in artificial intelligence has started to incorporate these inductive biases into the design of systems, leading to remarkable gains in performance.[5] In addition, progress has been made in automatically discovering the right inductive biases for a set of tasks, a technique known as *meta-learning*.[6]

Some of the approximation biases that I discussed are already incorporated into artificial intelligence systems. For example, Monte Carlo methods for approximate inference and random exploration for reinforcement learning are both widely used. Applications of efficient coding originated in telecommunications engineering. In the modern era, many consumer devices like cell phones typically send computationally expensive jobs to a server, thus minimizing the footprint on the device itself. Similarly, only a very small proportion of the enormous amount of data available on the internet is stored directly on a device at any given time. Nonetheless, the need for energy and space efficiency has become increasingly important as the complexity and scale of computations increase. If one wants to run a machine learning algorithm on videos collected by a phone, for example, simultaneously streaming the video, running the algorithm, and streaming back the results, one must reckon with limits on bandwidth and storage. It remains an open question whether the particular design solutions that the brain has evolved to meet its energy, bandwidth, and storage constraints are useful for engineering applications. The field of neuromorphic computing has taken this idea seriously, showing how chip designs that resemble neural circuits can achieve much greater energy efficiency.[7]

Implications for neuroscience

My focus in this book has been on cognition, and I give relatively short shrift to neuroscience. As a neuroscientist myself, I'm committed to the goal of integrating our understanding of mind and brain. However, I also believe that our current understanding of neuroscience is not yet ready to offer strong insights into the nature of intelligence.

This perspective will seem perverse to some readers, particularly those who have imbibed the recent zeitgeist of neuroscience-inspired artificial intelligence systems.[8] In my view, the direction of inspiration has flowed almost entirely in the opposite direction, from engineering to neuroscience. Many computational ideas about the brain have originated in the study of machine learning. Some well-known examples are the link between reward prediction errors and dopamine,[9] and modeling of the visual system using deep neural networks trained for computer vision tasks.[10]

These examples are instructive, because there is a certain folklore about the origins of machine learning in neuroscience. The inventors of reinforcement learning were inspired by animal learning in their development of algorithms that use errors to update reward predictions,[11] and early applications of deep neural networks to computer vision were inspired by the response properties of neurons in the visual cortex.[12] Thus, one could argue that the history goes precisely against my claim about directionality. Note, however, that these engineering ideas did not slavishly imitate biology, but rather recognized some underlying structure that biology is exploiting. The usefulness of reward prediction errors derives from the structure of certain sequential decision problems and the usefulness of certain visual response properties derives from certain invariance properties of scene geometry (e.g., object properties are mostly invariant when they undergo changes in position). The engineers who developed these algorithms understood this point clearly, because they deviated from biology whenever it was expedient. For example, deep neural networks typically propagate errors using a biologically implausible algorithm known as *backpropagation*. Neuroscientists have sought to find ways that the brain could approximate this algorithm.[13] But few engineers are using such biologically plausible variants in their artificial systems, since they inevitably perform worse than the original algorithm.

The lesson I draw from these examples is that the cognitive aspects of intelligence are fundamental for understanding neurobiological phenomena. We can learn much about the brain by understanding the structure of an information-processing problem. We can, for example, identify neurons that encode prediction errors and shift-invariant object properties, and understand how these neurons are arranged into more complex circuits to solve difficult problems. But we cannot discover the concepts of prediction error or invariant pattern recognition from "just looking" at what neurons do. We must come to the laboratory already equipped with these concepts if we have any hope of seeing their signatures in the data. For this reason, I believe that the major leaps in our understanding of neural computation arrive whenever we discover new computational principles expressed at the level of cognition.

Coda

The mathematician John von Neumann, who contributed to the development of computing (among many other things), was one of the first people to connect neuroscience to the nascent field of computer science. This naturally led him to think about what it would take to build a machine that thinks like people. E. T. Jaynes recounts the following story about a talk von Neumann gave at Princeton in 1948.[14] An audience member asked, "But, of course, a mere machine can't really *think*, can it?" To this von Neumann responded, "You insist that there is something that a machine cannot do. If you will tell me precisely what it is that a machine cannot do, then I can always make a machine which will do just that!"

I think many of us, like the audience member at von Neumann's talk, share the feeling that there is something inaccessible about human thought. In our daily lives, we have access to conscious thoughts that only skim the surface of what really goes on in our brains. A good portion of life is spent explaining ourselves—Why did I do that? What was I thinking? But it's a mistake to think that the limits of introspection define the limits of understanding. In this book, I have shown that we understand a great deal about why we think what we think and do what we do. We understand the answers with enough precision that we can implement them in computer programs that (to a first approximation) think the same thoughts and perform the same actions. Of course, the work is not done; we have a long way to go before we have human-like thinking machines. But we've come far enough to have optimism about the prospects of such an enterprise.

NOTES

Chapter 1. Introduction: Are we smart?

1. Ariely (2008); Marcus (2009); Kahneman (2011).

2. Knill and Richards (1996); Yuille and Kersten (2006); Todorov (2004); Griffiths and Tenenbaum (2006); Bahrami et al. (2010); Frank and Goodman (2012); Bogacz (2007); Oaksford and Chater (2007).

3. Nisbett and Ross (1980), p. xi.

4. Kleinmuntz (1990).

5. See Hakes and Sauer (2006) for a quantitative analysis supporting this argument.

6. Silver et al. (2016); Brown and Sandholm (2018); Moravcik et al. (2017); Silver et al. (2018); Ferrucci et al. (2010).

7. See Lake et al. (2017) and Davis and Marcus (2015) for extensive discussions.

8. Schank (1972).

9. Bar-Hillel (1964).

10. Winograd (1972), p. 33.

11. The most recent AI systems have not exceeded 70% on these types of problems, which are commonly known as *Winograd schemas* (Morgenstern et al., 2016).

12. Microsoft CaptionBot, https://www.captionbot.ai/.

13. These examples were not chosen to be adversarial—they were among the first images that I fed to the system. Also, note that these are not quirks unique to Microsoft's system. Similar issues afflict other systems.

14. Marcus (2009).

15. See Bowers and Davis (2012), Jones and Love (2011), and Marcus and Davis (2013).

16. I'm glossing over the fact that no one, not even the experimenter, has direct access to reality. It's better to think of an inferential system as reasoning about an "interface" with reality. See Hoffman (2019).

Chapter 2. Rational illusions

1. Rock and Kaufman (1962); Kaufman and Rock (1962). Note that the apparent distance explanation for the moon illusion is controversial; see, for example, Restle (1970).

2. Another answer, put forth by Hoffman (2019), is that veridical perception is not in general selected by evolution.

3. For our purposes, we don't necessarily need to think about the robot as a physical object; it could just be a computer looking at the world through a camera.

4. The phrase "the logic of perception" is borrowed from the title of Irvin Rock's book (Rock, 1983) of the same name.

5. See Hacking (1975) for a history of these two perspectives. Jaynes (2003) provides a foundational introduction to the epistemic perspective.

6. Gershman and Beck (2017); Doya et al. (2007).

7. Ma et al. (2006).

8. Buesing et al. (2011); Orbán et al. (2016).

9. Friston (2005); Huang and Rao (2011).

10. A decision rule is *admissible* if no other decision rule always leads to higher average reward. The admissibility of the Bayesian decision rule (with an appropriate choice of prior) is established by the *complete class theorems* (see Robert, 2007).

11. De Finetti (1931).

12. Lehman (1955); Kemeny (1955).

13. These ideas were first developed by Cox (1946). See Jaynes (2003) for an accessible introduction to these ideas.

14. Baum (2004).

15. Kanizsa (1955).

16. Gregory (1972).

17. Explaining away is sometimes referred to by the term "discounting" in cognitive psychology (Morris and Larrick, 1995), but this usage is different from how it has been used in some papers on perception (e.g., Kersten et al., 2004), where discounting refers to marginalization over nuisance variables.

18. Cornsweet (1970); Craik (1966); O'Brien (1959).

19. Warren et al. (1981); Battaglia et al. (2003).

20. Alais and Burr (2004).

21. Körding et al. (2007); Slutsky and Recanzone (2001); Jack and Thurlow (1973).

22. Körding et al. (2007).

23. Wallace et al. (2004).

24. Kahneman (2011); Gigerenzer and Gaissmaier (2011); Tversky and Kahneman (1974).

25. "Abuse of Iraqi POWs by GIs probed." CBS News, April 27, 2004.

26. Reviewed in Gilbert and Malone (1995).

27. "Red Cross saw "widespread abuse," BBC News, May 8, 2004.

28. Arendt (1964).

29. Arendt was also challenged on factual points, which I do not address here (see, for example, Haslam and Reicher, 2008).

30. Milgram (1974).

31. Haslam and Reicher (2008) have challenged this characterization, arguing that participants in Milgram's experiments did not completely cede their agency to the experimenters. Some wrestled with their consciences about the moral justifiability of their actions.

32. Jones and Harris (1967).

33. Morris and Larrick (1995).

34. Morris and Larrick (1995); Walker et al. (2015).

35. Ginzel et al. (1987).

36. Morris and Larrick (1995).

37. Juslin (1993); Griffin and Tversky (1992); Ferrell and McGoey (1980).
38. Moore and Healy (2008); Benoît and Dubra (2011).
39. Moore and Small (2007).

Chapter 3. Structure and origins of inductive bias

1. Michotte (1946).
2. Heider and Simmel (1944).
3. Gopnik and Meltzoff (1997); Gerstenberg and Tenenbaum (2017).
4. Pinker and Ullman (2002). It should be noted that the distinction between rules and exceptions has been extremely controversial, with some authors arguing that both regular and irregular verbs are represented in a single brain system (McClelland and Patterson, 2002).
5. Prasada and Pinker (1993).
6. Marcus et al. (1992); Maratsos (2000).
7. Chomsky (1986), p. 7.
8. Biederman (1987).
9. See Witkin and Tenenbaum (1983).
10. Lake et al. (2015).
11. Gershman et al. (2016). The origins of this idea trace back to Johansson (1950).
12. Duncker (1929).
13. Loomis and Nakayama (1973).
14. Nawrot and Sekuler (1990).
15. See Yamins and DiCarlo (2016) for an overview.
16. Scholl and Pylyshyn (1999).
17. Burke (1952); Flombaum and Scholl (2006).
18. Wertheimer (1912); Anstis (1980).
19. Burt and Sperling (1981); Kolers and Pomerantz (1971).
20. Valenza et al. (2006).
21. Zador (2019).
22. Gaier and Ha (2019).
23. Spelke (1990); Spelke et al. (1992).
24. Leslie and Keeble (1987).
25. Bonawitz et al. (2010). See also Goodman et al. (2011) for a computational view.
26. Piantadosi et al. (2018).
27. The idea of inferring an inductive bias goes back to the philosopher Nelson Goodman's concept of *overhypothesis* (Goodman, 1955), a term that has been incorporated into the cognitive science literature (e.g., Kemp et al., 2007). Goodman's treatment was not probabilistic, and one can formalize inductive bias learning without committing to a Bayesian framework (e.g., Baxter, 2000).
28. The classic studies of learning to learn were conducted by Harlow (1949).
29. McNamara (1982); Markman and Hutchinson (1984).
30. Landau et al. (1988).
31. Kemp et al. (2007).
32. Smith et al. (2002).

33. Polo (1918), p. 340.

34. Chater and Vitányi (2003).

35. A few representative examples: Gershman and Niv (2013); Gershman et al. (2017b); Austerweil and Griffiths (2013); Lau et al. (2018).

36. Kemp and Tenenbaum (2008).

37. Piantadosi and Jacobs (2016).

Chapter 4. Learning from others

1. Henrich (2017).

2. Henrich (2017), p. 5.

3. Asch (1956).

4. Bikhchandani et al. (1992); Anderson and Holt (1997).

5. See Bikhchandani et al. (1998) for an overview.

6. Welch (1992).

7. Farnsworth (2007).

8. Horner and Whiten (2005).

9. Whiten et al. (2016).

10. McGuigan et al. (2011).

11. Goodman et al. (2009).

12. Buchsbaum et al. (2011).

13. McGuigan et al. (2007); McGuigan and Whiten (2009).

14. Wimmer and Perner (1983); Perner et al. (1987).

15. Gopnik and Slaughter (1991).

16. See Keupp et al. (2018) for a review.

17. Keupp et al. (2016).

18. Schachner and Carey (2013).

19. Keupp et al. (2015).

20. Yaniv and Kleinberger (2000); Yaniv (2004).

21. Hawthorne-Madell and Goodman (2019).

22. Harvey and Fischer (1997); Yaniv (2004).

23. Bochner and Insko (1966); Aronson et al. (1963).

24. Ravazzolo and Røisland (2011); Hawthorne-Madell and Goodman (2019).

25. Yaniv (2004).

26. Bochner and Insko (1966); Aronson et al. (1963).

27. Birnbaum and Mellers (1983); Hawthorne-Madell and Goodman (2019).

28. Hawthorne-Madell and Goodman (2019).

29. Bhui and Gershman (2020).

30. Werch and Owen (2002).

Chapter 5. Good questions

1. See Nickerson (1998) for a review.

2. Klayman and Ha (1987).

3. Wason (1968); Johnson-Laird and Wason (1970).

4. Popper (1959).

5. Shannon and Weaver (1949).

6. Assuming the logarithm is base 2 (Cover and Thomas, 1991).

7. Divergence here is measured using the *relative entropy*, or Kullback-Leibler divergence, $D[P(X|Y)||P(X)] = \sum_x P(x|y) \log \frac{P(x|y)}{P(x)}$. The mutual information can then be expressed as: $I(X;Y) = \sum_y p(y)D[P(X|y)||P(X)]$.

8. Lindley et al. (1956).

9. Oaksford and Chater (1994).

10. See also Klayman and Ha (1987).

11. Kirby (1994); Wason and Green (1984).

12. Wason (1969).

13. Austerweil and Griffiths (2011a).

14. Navarro and Perfors (2011).

15. Hendrickson et al. (2016).

16. Cappelli et al. (2001).

17. Tversky and Shafir (1992).

18. Bennett et al. (2016); Cabrero et al. (2019).

19. Shipp et al. (2004).

20. Vohs et al. (2007).

21. Varga (2001).

22. Sweeny et al. (2010); Ajekigbe (1991).

23. Caplin and Leahy (2001).

24. Loewenstein (1987).

25. Pierson and Goodman (2014).

Chapter 6. How to never be wrong

1. This chapter is adapted from Gershman (2019a).

2. Popper (1959); Hempel (1966).

3. Harding (1976).

4. Popper (1959), p. 62.

5. Quine (1951), p. 38.

6. Gilovich (1991).

7. Dorling (1979); Earman (1992); Howson and Urbach (2006); Strevens (2001).

8. As a caveat, we should keep in mind that whether a particular intuitive theory satisfies these properties will naturally vary across domains and an individual's experience.

9. Duhem (1954), p. 187.

10. Quine (1951), pp. 42–43.

11. Lakatos (1976).

12. Kuhn (1962).

13. Grünbaum (1962); Laudan (1990).

14. Howson and Urbach (2006).

15. Dorling (1979); Earman (1992); Howson and Urbach (2006); Strevens (2001).

16. Fitelson and Waterman (2005); Mayo (1997).

17. Strevens (2001) notes that this expression does not hold if the data affect the posterior in ways other than falsifying the conjunct *ha*, although such scenarios are probably rare.

18. We can think of this conditional prior as specifying strength of belief in an auxiliary given that one already believes a particular central hypothesis. In other words, it assumes that different central hypotheses invoke different distributions over auxiliaries. This seems intuitive insofar as auxiliaries will tend to be highly theory-specific (you don't hypothesize auxiliaries about baseball when contemplating cosmology).

19. This is not to deny that some forms of motivated reasoning exist (see Kunda, 1990), but only to assert particular ways in which robustness to disconfirmation arises from rational inference.

20. See Gershman (2019a) for more details.

21. Klayman and Ha (1987).

22. Oaksford and Chater (1994).

23. Hendrickson et al. (2016); McKenzie et al. (2001).

24. Mckenzie and Mikkelsen (2000); McKenzie and Mikkelsen (2007).

25. Perfors and Navarro (2009).

26. Note that there are a number of reasons why people might generate sparse hypotheses besides having a sparse prior, such as computational limits (cf. Dasgupta et al., 2017).

27. Olshausen and Field (1996).

28. Hromádka et al. (2008).

29. Poo and Isaacson (2009).

30. Austerweil and Griffiths (2013); Biederman (1987).

31. Navarro and Perfors (2011).

32. Rosch (1978).

33. Schulz and Sommerville (2006); Muentener and Schulz (2014); Wu et al. (2015); Saxe et al. (2005); Buchanan and Sobel (2011).

34. Schulz and Sommerville (2006).

35. Saxe et al. (2005).

36. Mayrhofer and Waldmann (2015); Frosch and Johnson-Laird (2011). Some evidence suggests that people can adaptively determine which causal theory (deterministic or probabilistic) is most suitable for a given domain (Griffiths and Tenenbaum, 2009).

37. Austerweil and Griffiths (2011b).

38. Lu et al. (2008).

39. Yeung and Griffiths (2015) presented empirical evidence favoring a preference for (near) determinism but not sparsity, though other experiments have suggested that both sparsity and determinism are required to explain human causal inferences (Powell et al., 2016).

40. Buchanan et al. (2010).

41. Lakatos (1976).

42. Kunda (1990).

43. Indeed, there has been a vigorous debate in psychology about the validity of Bayesian rationality as a model of human cognition (Jones and Love, 2011). Here I am merely asking the reader to consider the conditional claim that *if* people are Bayesian with sparse and deterministic intuitive theories, then they would exhibit robustness to disconfirmation.

44. Kuhn (1962), p. 113.

45. Stratton (1897).

46. Kuhn (1962), pp. 111–112.

47. Helmholtz (1867); Gregory (1970); Rock (1983).

48. It is important to distinguish this view from the stronger thesis that no theory-neutral stage of perceptual analysis exists (e.g., Churchland, 1979). As pointed out by Fodor (1984), we can accept that the semantic interpretation of percepts is theory-dependent without abandoning the possibility that there are *some* cognitively impenetrable aspects of perception.

49. See Bovens and Hartmann (2002) for a detailed analysis of how beliefs about the unreliability of measurement instruments affects reasoning about auxiliary hypotheses.

50. How can this debunking strategy succeed when theorists can produce new auxiliary hypotheses ad infinitum? The Bayesian analysis makes provision for this: new auxiliaries will only be considered if they have appreciable probability, $P(a|h)$, relative to the prior, $P(h)$.

51. Leiserowitz et al. (2013).

52. Lord et al. (1979).

53. (E.g., Nisbett and Ross, 1980; Kunda, 1990).

54. Jern et al. (2014); Cook and Lewandowsky (2016); Koehler (1993); Jaynes (2003).

55. Sharot (2011).

56. The generality of this effect has been the subject of controversy, with some authors, like Shah et al. (2016), finding no evidence for an optimism bias. However, these null results have themselves been controversial: correcting confounds in the methodology (Garrett and Sharot, 2017), and using model-based estimation techniques (Kuzmanovic and Rigoux, 2017), have indicated a robust optimism bias.

57. Stankevicius et al. (2014).

58. Eil and Rao (2011); Sharot and Garrett (2016).

59. Eil and Rao (2011).

60. Sharot et al. (2011); Korn et al. (2012); Lefebvre et al. (2017).

61. Eil and Rao (2011); Köszegi (2006).

62. Gibson and Sanbonmatsu (2004).

63. Gilovich (1983); Gilovich and Douglas (1986).

64. Gervais and Odean (2001).

65. Gino et al. (2016).

66. Chance et al. (2011).

67. Dorfman et al. (2019).

68. Huys and Dayan (2009).

69. Harris (1996); Weinstein (1980).

70. Alloy et al. (1981).

71. Moore and Fresco (2012).

72. Korn et al. (2014).

73. Seligman (1975); Huys and Dayan (2009); Abramson et al. (1978).

74. Langer (1975).

75. Ladouceur and Sévigny (2005).

76. See also Gilovich and Douglas (1986); Fernandez-Duque and Wifall (2007).

77. Strohminger et al. (2017).

78. Haslam et al. (2004).

79. Christy et al. (2017).

80. Molouki and Bartels (2017); De Freitas et al. (2017).

81. Newman et al. (2014).

82. De Freitas et al. (2018).

83. Newman et al. (2015).

84. Bench et al. (2015); De Freitas et al. (2017).

85. Hewstone (1994).

86. Weber and Crocker (1983); Hewstone (1994); Johnston and Hewstone (1992).

87. Weber and Crocker (1983).

88. See also Johnston and Hewstone (1992).

89. Van Rooy et al. (2003).

90. Queller and Smith (2002).

91. Bouton (2004).

92. Rescorla (2004).

93. Weber and Crocker (1983); Johnston and Hewstone (1992).

94. Chelonis et al. (1999); Gunther et al. (1998).

95. Queller and Smith (2002).

96. Gershman et al. (2013). See also Gershman et al. (2014) for related findings in humans.

97. Gershman et al. (2017a).

98. Sunstein and Vermeule (2009), p. 205.

99. Sunstein and Vermeule (2009), p. 207.

100. Chater and Loewenstein (2016).

101. MacKay (2003).

102. Lewandowsky et al. (2013); Goertzel (1994).

103. Swinburne (2004).

104. Boyer (2003).

105. Earman (2000); Swinburne (1970).

106. Howson (2011).

107. This point is closely related to the position known as *skeptical theism* (McBrayer, 2010), which argues that our inability to apprehend God's reasons for certain events (e.g., evil) does not justify the claim that no such reasons exist. This position undercuts inductive arguments against the existence of God that rely on the premise that no reasons exist for certain events.

108. Carey (2009); Gopnik (2012).

109. Gopnik and Wellman (1992).

110. Vosniadou and Brewer (1992).

111. Champagne et al. (1985).

112. Karmiloff-Smith and Inhelder (1975).

113. Schulz et al. (2008).

114. See also Schulz and Sommerville (2006); Saxe et al. (2005).

115. Chinn and Brewer (1993), p. 2.

116. Marcus and Davis (2013); Bowers and Davis (2012).

117. Grether (1980); Evans et al. (2002).

118. See Dasgupta et al. (2017).

119. Quine (1951).

Chapter 7. Seeing patterns

1. Gilovich (1991).

2. Shaffer and Jadwiszczok (2016).

3. Diaconis (1978); Wagenmakers et al. (2011). For a positive reading of the evidence, see Cardeña (2018).

4. Griffiths and Tenenbaum (2007).

5. Kao and Wasserman (1993); Mandel and Lehman (1998); Schustack and Sternberg (1981).

6. McKenzie and Mikkelsen (2007); Anderson (1990).

7. McKenzie and Mikkelsen (2007).

8. Alloy and Tabachnik (1984).

9. Chapman and Chapman (1967).

10. Peterson (1980).

11. Clotfelter and Cook (1993).

12. Braun and Schmidt (2015).

13. Chen et al. (2016).

14. Gilovich et al. (1985); Koehler and Conley (2003). It should be noted, however, that the hot-hand fallacy has been the subject of intense controversy, because some have argued that hot hands might really exist (Miller and Sanjurjo, 2018).

15. Camerer (1989).

16. Rabin and Vayanos (2010).

17. Asparouhova et al. (2009).

18. Suetens et al. (2016).

19. Burns and Corpus (2004); Asparouhova et al. (2009).

Chapter 8. Are we consistent?

1. Von Neumann and Morgenstern (1944).

2. See, for example, Kahneman (2011).

3. Huber et al. (1982).

4. Simonson (1989).

5. Tversky (1972).

6. Wernerfelt (1995); Prelec et al. (1997); Shenoy and Yu (2013).

7. Shenoy and Yu (2013).

8. Hsee (1998).

9. List (2002).

10. Sher and McKenzie (2014).

11. See Lichtenstein and Slovic (2006) for an overview.

12. Nisbett and Wilson (1977), p. 244.

13. Brehm (1956).

14. Bem and McConnell (1970); Goethals and Reckman (1973).

15. Festinger and Carlsmith (1959).

16. Johansson et al. (2005); Hall et al. (2010).

17. Johansson et al. (2014).

18. Simons and Levin (1998).

19. Strandberg et al. (2018).

20. Cushman (2019).

21. Prendergast (1999).

22. Lazear (2000).

23. Kool et al. (2010, 2017); Westbrook et al. (2013).

24. Deci (1971).

25. Deci et al. (1999). It should be noted, however, that some authors dispute the ubiquity of the undermining effect, and suggest that it can be explained by other factors (Cameron et al., 2001).

26. Gneezy and Rustichini (2000b).

27. Frey and Oberholzer-Gee (1997).

28. Titmuss (1970).

29. Kohn (1993).

30. Gneezy and Rustichini (2000a).

31. Fehr and Gächter (2001).

32. Frey and Jegen (2001); Bénabou and Tirole (2003).

33. Lepper et al. (1973). These authors refer to "crowding out" as the *overjustification effect*.

34. See the model of "self-signaling" in Bodner and Prelec (2003).

35. Berglas and Jones (1978). See also Tucker et al. (1981) for similar results with alcohol consumption.

36. Gershman et al. (2017b).

37. See Bénabou and Tirole (2004, 2006).

38. Freedman and Fraser (1966).

39. Monin and Miller (2001).

40. Merritt et al. (2012).

41. Bénabou and Tirole (2004).

42. Burger (1999).

43. Freedman and Fraser (1966).

44. Gneezy et al. (2012).

45. Strotz (1955).

46. Schelling (1978), p. 290.

47. Milkman et al. (2010).

48. Thaler and Shefrin (1981). See also Fudenberg and Levine (2006).

49. See Ainslie (1992) for an extensive discussion.

50. Camerer et al. (1997).

51. Bénabou and Tirole (2004), p. 855.

52. These ideas are rooted in the classic social psychological treatise on "self-perception" by Bem (1972).

53. Quattrone and Tversky (1984).

54. For a broad survey of the multiple self idea, see the essays collected in Elster (1987).

55. Bartels and Rips (2010); Bartels and Urminsky (2011).

56. Bartels et al. (2013).

57. Van Gelder et al. (2013, 2015); Hershfield et al. (2011); Rutchick et al. (2018).

58. Parfit (1984). See also Paul (2014).

59. Stanley et al. (2019).

60. Frankfurt (1971).

61. George (2009).

62. Parfit (1984), p. 327.

63. Greene et al. (2001).

64. Tetlock et al. (2000). See Bénabou and Tirole (2011) for a self-inference model of these and related phenomena.

65. Wilson (2004).

Chapter 9. Celestial teapots and flying spaghetti monsters

1. "Austrian driver allowed 'pastafarian' headgear photo," BBC News, July 14, 2011.

2. "Pastafarian man wins right to wear colander on his head in driving licence photo," *Independent*, January 14, 2016.

3. From Russell's 1952 unpublished essay, "Is There a God?" in *The Collected Papers of Bertrand Russell, Vol. 11: Last Philosophical Testament, 1943–68* (1997), ed. John G. Slater, Routledge, pp. 542–548.

4. Hahn and Oaksford (2007).

5. Bowerman (1988).

6. Tsividis et al. (2017).

7. Oaksford and Hahn (2004).

8. Walton (1996). In computer science, epistemic closure is known as the *closed-world assumption* (Reiter, 1980).

9. For experimental evidence, see Hahn et al. (2005).

10. Grice (1989).

11. Harris et al. (2013).

12. Godwin, Mike (October 1994), "Meme, counter-meme," *Wired*.

13. Strauss (1953).

14. Bhatia and Oaksford (2015).

15. Harris et al. (2012).

16. Johnson-Laird (2006).

17. Mercier and Sperber (2017).

Chapter 10. The frugal brain

1. Note, however, that the brain, unlike conventional computers, is massively parallel, and this parallelism can be harnessed to perform certain computations more efficiently (Nelson and Bower, 1990).

2. This claim is sometimes referred to as *computational rationality* (Gershman et al., 2015), or *resource rationality* (Lieder and Griffiths, 2019).

3. One might argue that visual information arrives in digital form (photons), but photoreceptors in the retina are tuned to particular wavelengths—a continuous quantity.

4. Shannon's source coding theorem (Shannon and Weaver, 1949).

5. Shannon's noisy channel coding theorem (Shannon and Weaver, 1949).

6. This follows from the fact that coding a random variable by its rank (proportional to the cumulative distribution function) produces a uniformly distributed code, a technique known as the *probability integral transform*. See Bhui and Gershman (2018).

7. Konkle and Oliva (2011).

8. Voss and Clarke (1975).

9. Knudsen (1923).

10. Voss and Clarke (1975).

11. Knudsen (1923).

12. Fechner (1860).

13. Stevens (1957).

14. Knudsen (1923).

15. Werner and Mountcastle (1965); Nieder and Miller (2003).

16. See, for example, Schroeder (1991); Newman (2005).

17. Zipf (1949). The word-frequency distribution is the origin of what is now known as *Zipf's law*.

18. Mora and Bialek (2011).

19. Aitchison et al. (2016).

20. Mandelbrot (1953).

21. Srinivasan et al. (1982).

22. Brady and Alvarez (2011).

23. Ben-Shalom and Ganel (2012).

24. Wu and Chen (2018).

25. Wei and Stocker (2015).

26. Golomb (2015); Chunharas et al. (2019).

27. Stewart et al. (2006).

28. Vul et al. (2014).

29. Bhui and Gershman (2018).

30. See Bhui and Gershman (2018) for more details. The idea was based on an earlier model developed by Brown and Matthews (2011).

31. Pollack (1952).

32. Pollack (1953).

33. Parducci (1965).

34. Bhui and Gershman (2018).

35. Parducci and Perrett (1971).

36. Parducci and Wedell (1986).

37. Brown et al. (2008).

38. Luttmer (2005).

39. Rutledge et al. (2014).

40. Mussweiler (2003).

41. Wedell et al. (1987); Furl (2016). The rank effect in Wedell et al. (1987) is subtle, because it only appears with respect to faces appearing on other trials; the effect of faces on the current trial is *assimilative*, whereby attractive decoys increase the attractiveness of the target.

42. Morse and Gergen (1970).

43. Chang et al. (2019).

44. Kahneman and Tversky (1979); Köszegi (2006).

45. Mellers et al. (1997).

46. Kahneman and Tversky (1979).

47. Kahneman et al. (1990).

48. Coase (1960).

49. Shefrin and Statman (1985).

50. Odean (1998). In fact, it has been argued that stock market momentum is actually *caused* by "disposition investors" (Grinblatt and Han, 2005). The basic idea is that the disposition effect causes under-reaction to public information, thereby creating a discrepancy between a stock's current and rational valuation: high-performing stocks are undervalued (and hence should increase in price over time), and low-performing stocks are overvalued (and hence should decrease in price over time).

51. Mehra and Prescott (1985).

52. Benartzi and Thaler (1995).

53. Gneezy and Potters (1997); Thaler et al. (1997).

54. Stewart et al. (2006).

55. Stewart et al. (2014).

56. Walasek and Stewart (2015).

57. Bhui and Gershman (2018).

58. Louie et al. (2013).

59. Janiszewski and Lichtenstein (1999).

Chapter 11. Language design

1. Okrent (2009).

2. See Gibson et al. (2019) for a general overview of this approach to language. The role of effort in language design goes back to Zipf (1949).

3. Ben Zimmer, "On language: Crash blossoms," *New York Times Magazine*, January 27, 2010.

4. Cowan (2016), p. 3144.

5. Don Oldenbeurg, "Lojban: An artificial language seeks status as the lingua franca: Communications: There have been several attempts to create a universal tongue. This is one of the latest," *Los Angeles Times*, November 17, 1989.

6. Piantadosi et al. (2012).

7. Chomsky (2002), p. 107.

8. The Portable Benjamin Franklin (2005), ed. L. Ziff, New York: Penguin Books, p. 375.

9. Kliegl et al. (2004).

10. Piantadosi et al. (2012).

11. See Kemp et al. (2018) for an overview of this perspective.

12. Berlin and Kay (1969).

13. Regier and Kay (2009).

14. Roberson et al. (2005).

15. Goldstone and Hendrickson (2010).

16. Roberson et al. (2005).

17. Dowman (2007); Regier et al. (2015).

18. Gibson et al. (2017).

19. Boas (1911).

20. Regier et al. (2016).

21. Lupyan and Dale (2016).

22. Regier and Kay (2009).

23. Gell-Mann and Ruhlen (2011).

24. Goldin-Meadow et al. (2008).

25. Gibson et al. (2013b).

26. Gibson et al. (2013a).

27. Futrell et al. (2015).

28. Gibson et al. (2013b).

29. Schmidt et al. (2009).

30. Scontras et al. (2017).

31. Qian and Jaeger (2012).

32. Hahn et al. (2018).

33. Hauser et al. (2002).

34. Brighton et al. (2005).

35. Chomsky (1965).

36. Nowak et al. (2002).

37. Hsu and Chater (2010); Hsu et al. (2011).

38. McAllester (1999).

39. Kirby et al. (2007).

40. Kirby et al. (2015).

41. Kirby et al. (2008).

42. Kirby et al. (2015).

Chapter 12. The uses of randomness

1. Eckhardt (1987), p. 131.

2. Dyson (2012), p. 191.

3. Robert and Casella (2013).

4. See Glimcher (2005) for a review of the evidence.

5. Tolhurst et al. (1983).

6. Laughlin (2001).

7. Eagle (2005).

8. Mainen and Sejnowski (1995).

9. Deneve (2008).

10. Lee and Yuille (2006).

11. Kapadia et al. (2000).

12. Buesing et al. (2011); Pecevski et al. (2011).

13. Berkes et al. (2011).

14. Orbán et al. (2016).

15. Churchland et al. (2010); Orbán et al. (2016).

16. Gershman et al. (2012).

17. Knapen et al. (2007); Burke et al. (1999).

18. O'Shea et al. (1997).

19. Wilson et al. (2001).

20. Lee et al. (2005).

21. Vul et al. (2014).

22. Nosofsky et al. (1994).

23. Goodman et al. (2008).

24. Vul et al. (2014).

25. Griffiths and Tenenbaum (2006).

26. Lewandowsky et al. (2009); Mozer et al. (2008).

27. Denison et al. (2013).

28. Thaker et al. (2017); Dasgupta et al. (2017).

29. Vul et al. (2014).

30. Galinsky and Mussweiler (2001).

31. Englich et al. (2006).

32. Lieder et al. (2018a).

33. Epley and Gilovich (2006).

34. Epley and Gilovich (2005); Lieder et al. (2018b).

35. Dasgupta et al. (2017).

36. See Schulz et al. (2019) for a real-world study related to this example.

37. P. Whittle, commentary, in Gittins (1979), p. 165.

38. Gershman (2019b).

39. Tomov et al. (2019).

40. Agrawal and Goyal (2012).

41. Cesa-Bianchi et al. (2017).

42. Wagenaar (1972).

43. Rapoport and Budescu (1992).

44. Chiappori et al. (2002); Palacios-Huerta (2003).

45. Walker and Wooders (2001).

46. Driver and Humphries (1988).

47. Chance (1957).

48. Roeder (1962).

49. Driver and Humphries (1988).

50. Kennedy and Booth (1963).

51. Driver and Humphries (1988).

52. Miller (1997).

53. See also Icard (2019) for another rationalization of randomness based on limited memory, which is a bit too technical to expostulate here.

Chapter 13. Conclusion: What makes us smart

1. See Tauber et al. (2017) for further discussion of this point.

2. Lake et al. (2017).

3. Tsividis et al. (2017).

4. Dubey et al. (2018).

5. See Lake et al. (2017) for a discussion of some of this work.

6. Botvinick et al. (2019).

7. Monroe (2014).

8. Hassabis et al. (2017).

9. Schultz et al. (1997).

10. Yamins and DiCarlo (2016).

11. Sutton and Barto (2018).

12. Fukushima (1980).

13. Whittington and Bogacz (2019).

14. Jaynes (2003), p. 7.

BIBLIOGRAPHY

Abramson, L. Y., Seligman, M. E., and Teasdale, J. D. (1978). Learned helplessness in humans: Critique and reformulation. *Journal of Abnormal Psychology*, 87:49–74.

Adelson, E. H. (2000). Lightness perception and lightness illusions. In Gazzaniga, M., editor, *The New Cognitive Neurosciences*, 2nd ed., pages 339–351. MIT Press.

Agrawal, S. and Goyal, N. (2012). Analysis of Thompson sampling for the multi-armed bandit problem. In *Proceedings of the 25th Annual Conference on Learning Theory*, pages 39.1–39.26.

Ainslie, G. (1992). *Picoeconomics: The Strategic Interaction of Successive Motivational States within the Person*. Cambridge University Press.

Aitchison, L., Corradi, N., and Latham, P. E. (2016). Zipf's law arises naturally when there are underlying, unobserved variables. *PLoS Computational Biology*, 12:e1005110.

Ajekigbe, A. (1991). Fear of mastectomy: The most common factor responsible for late presentation of carcinoma of the breast in Nigeria. *Clinical Oncology*, 3:78–80.

Alais, D. and Burr, D. (2004). The ventriloquist effect results from near-optimal bimodal integration. *Current Biology*, 14:257–262.

Alloy, L. and Tabachnik, N. (1984). Assessment of covariation by humans and animals. *Psychological Review*, 91:112–149.

Alloy, L. B., Abramson, L. Y., and Viscusi, D. (1981). Induced mood and the illusion of control. *Journal of Personality and Social Psychology*, 41:1129–1140.

Anderson, J. R. (1990). *The Adaptive Character of Thought*. Psychology Press.

Anderson, L. R. and Holt, C. A. (1997). Information cascades in the laboratory. *The American Economic Review*, 87:847–862.

Anstis, S. I. (1980). The perception of apparent movement. *Philosophical Transactions of the Royal Society of London. B, Biological Sciences*, 290:153–168.

Arendt, H. (1964). *Eichmann in Jerusalem*. Penguin.

Ariely, D. (2008). *Predictably Irrational*. HarperCollins.

Aronson, E., Turner, J., and Carlsmith, J. (1963). Communicator credibility and communication discrepancy as determinants of opinion change. *The Journal of Abnormal and Social Psychology*, 67:31–36.

Asch, S. (1956). Studies of independence and conformity: I. A minority of one against a unanimous majority. *Psychological Monographs: General and Applied*, 70:1–70.

Asparouhova, E., Hertzel, M., and Lemmon, M. (2009). Inference from streaks in random outcomes: Experimental evidence on beliefs in regime shifting and the law of small numbers. *Management Science*, 55:1766–1782.

Austerweil, J. L. and Griffiths, T. L. (2011a). A rational model of the effects of distributional information on feature learning. *Cognitive Psychology*, 63:173–209.

Austerweil, J. L. and Griffiths, T. L. (2011b). Seeking confirmation is rational for deterministic hypotheses. *Cognitive Science*, 35:499–526.

Austerweil, J. L. and Griffiths, T. L. (2013). A nonparametric Bayesian framework for constructing flexible feature representations. *Psychological Review*, 120:817–851.

Bahrami, B., Olsen, K., Latham, P. E., Roepstorff, A., Rees, G., and Frith, C. D. (2010). Optimally interacting minds. *Science*, 329:1081–1085.

Bar-Hillel, Y. (1964). *Language and Information*. Addison-Wesley.

Bartels, D. and Rips, L. (2010). Psychological connectedness and intertemporal choice. *Journal of Experimental Psychology: General*, 139:49–69.

Bartels, D. M., Kvaran, T., and Nichols, S. (2013). Selfless giving. *Cognition*, 129:392–403.

Bartels, D. M. and Urminsky, O. (2011). On intertemporal selfishness: How the perceived instability of identity underlies impatient consumption. *Journal of Consumer Research*, 38:182–198.

Battaglia, P. W., Jacobs, R. A., and Aslin, R. N. (2003). Bayesian integration of visual and auditory signals for spatial localization. *Journal of the Optical Society of America A*, 20:1391–1397.

Baum, E. B. (2004). *What Is Thought?* MIT Press.

Baxter, J. (2000). A model of inductive bias learning. *Journal of Artificial Intelligence Research*, 12:149–198.

Bem, D. and McConnell, H. (1970). Testing the self-perception explanation of dissonance phenomena. *Journal of Personality and Social Psychology*, 14:23–31.

Bem, D. J. (1972). Self-perception theory. In *Advances in Experimental Social Psychology*, volume 6, pages 1–62. Elsevier.

Ben-Shalom, A. and Ganel, T. (2012). Object representations in visual memory: Evidence from visual illusions. *Journal of Vision*, 12:15–15.

Bénabou, R. and Tirole, J. (2003). Intrinsic and extrinsic motivation. *The Review of Economic Studies*, 70:489–520.

Bénabou, R. and Tirole, J. (2004). Willpower and personal rules. *Journal of Political Economy*, 112:848–886.

Bénabou, R. and Tirole, J. (2006). Incentives and prosocial behavior. *American Economic Review*, 96:1652–1678.

Bénabou, R. and Tirole, J. (2011). Identity, morals, and taboos: Beliefs as assets. *The Quarterly Journal of Economics*, 126:805–855.

Benartzi, S. and Thaler, R. H. (1995). Myopic loss aversion and the equity premium puzzle. *The Quarterly Journal of Economics*, 110:73–92.

Bench, S. W., Schlegel, R. J., Davis, W. E., and Vess, M. (2015). Thinking about change in the self and others: The role of self-discovery metaphors and the true self. *Social Cognition*, 33:169–185.

Bennett, D., Bode, S., Brydevall, M., Warren, H., and Murawski, C. (2016). Intrinsic valuation of information in decision making under uncertainty. *PLoS Computational Biology*, 12:e1005020.

Benoît, J.-P. and Dubra, J. (2011). Apparent overconfidence. *Econometrica*, 79:1591–1625.

Berglas, S. and Jones, E. (1978). Drug choice as a self-handicapping strategy in response to noncontingent success. *Journal of Personality and Social Psychology*, 36:405–417.

Berkes, P., Orbán, G., Lengyel, M., and Fiser, J. (2011). Spontaneous cortical activity reveals hallmarks of an optimal internal model of the environment. *Science*, 331:83–87.

Berlin, B. and Kay, P. (1969). *Basic Color Terms: Their Universality and Evolution.* University of California Press.

Bhatia, J.-S. and Oaksford, M. (2015). Discounting testimony with the argument ad hominem and a Bayesian congruent prior model. *Journal of Experimental Psychology: Learning, Memory, and Cognition,* 41:1548–1559.

Bhui, R. and Gershman, S. J. (2018). Decision by sampling implements efficient coding of psychoeconomic functions. *Psychological Review,* 125:985–1001.

Bhui, R. and Gershman, S. J. (2020). Paradoxical effects of persuasive messages. *Decision,* 7:239–258.

Biederman, I. (1987). Recognition-by-components: A theory of human image understanding. *Psychological Review,* 94:115–147.

Biederman, I. (1995). Visual object recognition. IIn Kosslyn, M. and Osherson, D. N., editors, *An Invitation to Cognitive Science: Visual Cognition,* Vol. 2, 2nd ed., pp. 121–165. MIT Press.

Bikhchandani, S., Hirshleifer, D., and Welch, I. (1992). A theory of fads, fashion, custom, and cultural change as informational cascades. *Journal of Political Economy,* 100:992–1026.

Bikhchandani, S., Hirshleifer, D., and Welch, I. (1998). Learning from the behavior of others: Conformity, fads, and informational cascades. *Journal of Economic Perspectives,* 12:151–170.

Birnbaum, M. and Mellers, B. (1983). Bayesian inference: Combining base rates with opinions of sources who vary in credibility. *Journal of Personality and Social Psychology,* 45:792–804.

Boas, F. (1911). *Introduction to Handbook of American Indian Languages.* Number 677. US Government Printing Office.

Bochner, S. and Insko, C. (1966). Communicator discrepancy, source credibility, and opinion change. *Journal of Personality and Social Psychology,* 4:614–621.

Bodner, R. and Prelec, D. (2003). Self-signaling and diagnostic utility in everyday decision making. In *The Psychology of Economic Decisions,* pages 105–26. Oxford University Press.

Bogacz, R. (2007). Optimal decision-making theories: Linking neurobiology with behaviour. *Trends in Cognitive Sciences,* 11:118–125.

Bonawitz, E. B., Ferranti, D., Saxe, R., Gopnik, A., Meltzoff, A. N., Woodward, J., and Schulz, L. E. (2010). Just do it? Investigating the gap between prediction and action in toddlers' causal inferences. *Cognition,* 115:104–117.

Botvinick, M., Ritter, S., Wang, J. X., Kurth-Nelson, Z., Blundell, C., and Hassabis, D. (2019). Reinforcement learning, fast and slow. *Trends in Cognitive Sciences,* 23:408–422.

Bouton, M. E. (2004). Context and behavioral processes in extinction. *Learning & Memory,* 11:485–494.

Bovens, L. and Hartmann, S. (2002). Bayesian networks and the problem of unreliable instruments. *Philosophy of Science,* 69:29–72.

Bowerman, M. (1988). The 'no negative evidence' problem: How do children avoid constructing an overly general grammar? In Hawkins, J., editor, *Explaining Language Universals,* pages 73–101. Basil Blackwell.

Bowers, J. S. and Davis, C. J. (2012). Bayesian just-so stories in psychology and neuroscience. *Psychological Bulletin,* 138(3):389–414.

Boyer, P. (2003). Religious thought and behaviour as by-products of brain function. *Trends in Cognitive Sciences,* 7:119–124.

Brady, T. F. and Alvarez, G. A. (2011). Hierarchical encoding in visual working memory: Ensemble statistics bias memory for individual items. *Psychological Science,* 22:384–392.

Braun, S. and Schmidt, U. (2015). The gambler's fallacy in penalty shootouts. *Current Biology*, 25:R597–R598.

Brehm, J. (1956). Postdecision changes in the desirability of alternatives. *The Journal of Abnormal and Social Psychology*, 52:384–389.

Brighton, H., Smith, K., and Kirby, S. (2005). Language as an evolutionary system. *Physics of Life Reviews*, 2:177–226.

Brown, G. D., Gardner, J., Oswald, A. J., and Qian, J. (2008). Does wage rank affect employees' well-being? *Industrial Relations: A Journal of Economy and Society*, 47:355–389.

Brown, G. D. and Matthews, W. J. (2011). Decision by sampling and memory distinctiveness: Range effects from rank-based models of judgment and choice. *Frontiers in Psychology*, 2:299.

Brown, N. and Sandholm, T. (2018). Superhuman AI for heads-up no-limit poker: Libratus beats top professionals. *Science*, 359:418–424.

Buchanan, D. W. and Sobel, D. M. (2011). Children posit hidden causes to explain causal variability. In *Proceedings of the 33rd Annual Conference of the Cognitive Science Society*.

Buchanan, D. W., Tenenbaum, J. B., and Sobel, D. M. (2010). Edge replacement and nonindependence in causation. In *Proceedings of the 32nd Annual Conference of the Cognitive Science Society*, pages 919–924.

Buchsbaum, D., Gopnik, A., Griffiths, T. L., and Shafto, P. (2011). Children's imitation of causal action sequences is influenced by statistical and pedagogical evidence. *Cognition*, 120: 331–340.

Buesing, L., Bill, J., Nessler, B., and Maass, W. (2011). Neural dynamics as sampling: A model for stochastic computation in recurrent networks of spiking neurons. *PLoS Computational Biology*, 7:e1002211.

Burger, J. M. (1999). The foot-in-the-door compliance procedure: A multiple-process analysis and review. *Personality and Social Psychology Review*, 3:303–325.

Burke, D., Alais, D., and Wenderoth, P. (1999). Determinants of fusion of dichoptically presented orthogonal gratings. *Perception*, 28:73–88.

Burke, L. (1952). On the tunnel effect. *Quarterly Journal of Experimental Psychology*, 4: 121–138.

Burns, B. D. and Corpus, B. (2004). Randomness and inductions from streaks: "Gambler's fallacy" versus "hot hand." *Psychonomic Bulletin & Review*, 11:179–184.

Burt, P. and Sperling, G. (1981). Time, distance, and feature trade-offs in visual apparent motion. *Psychological Review*, 88:171–195.

Cabrero, J. M. R., Zhu, J.-Q., and Ludvig, E. A. (2019). Costly curiosity: People pay a price to resolve an uncertain gamble early. *Behavioural Processes*, 160:20–25.

Camerer, C., Babcock, L., Loewenstein, G., and Thaler, R. (1997). Labor supply of New York City cabdrivers: One day at a time. *The Quarterly Journal of Economics*, 112:407–441.

Camerer, C. F. (1989). Does the basketball market believe in the 'hot hand,'? *American Economic Review*, 79:1257–1261.

Cameron, J., Banko, K. M., and Pierce, W. D. (2001). Pervasive negative effects of rewards on intrinsic motivation: The myth continues. *The Behavior Analyst*, 24:1–44.

Caplin, A. and Leahy, J. (2001). Psychological expected utility theory and anticipatory feelings. *The Quarterly Journal of Economics*, 116:55–79.

Cappelli, M., Surh, L., Humphreys, L., Verma, S., Logan, D., Hunter, A., and Allanson, J. (2001). Measuring women's preferences for breast cancer treatments and BRCA1/BRCA2 testing. *Quality of Life Research*, 10:595–607.

Cardeña, E. (2018). The experimental evidence for parapsychological phenomena. *American Psychologist*, 73:663–677.

Carey, S. (2009). *The Origin of Concepts*. Oxford University Press.

Cesa-Bianchi, N., Gentile, C., Lugosi, G., and Neu, G. (2017). Boltzmann exploration done right. In *Advances in Neural Information Processing Systems*, pages 6284–6293.

Champagne, A., Gunstone, R. F., and Klopfer, L. E. (1985). Instructional consequences of students' knowledge about physical phenomena. In West, L. and Pines, A., editors, *Cognitive Structure and Conceptual Change*, pages 61–68. Academic Press.

Chance, M. (1957). The role of convulsions in behavior. *Behavioral Science*, 2:30–45.

Chance, Z., Norton, M. I., Gino, F., and Ariely, D. (2011). Temporal view of the costs and benefits of self-deception. *Proceedings of the National Academy of Sciences*, 108:15655–15659.

Chang, L. W., Gershman, S. J., and Cikara, M. (2019). Comparing value coding models of context-dependence in social choice. *Journal of Experimental Social Psychology*, 85:103847.

Chapman, L. and Chapman, J. (1967). Genesis of popular but erroneous psychodiagnostic observations. *Journal of Abnormal Psychology*, 72:193–204.

Chater, N. and Loewenstein, G. (2016). The under-appreciated drive for sense-making. *Journal of Economic Behavior & Organization*, 126:137–154.

Chater, N. and Vitányi, P. (2003). Simplicity: A unifying principle in cognitive science? *Trends in Cognitive Sciences*, 7:19–22.

Chelonis, J. J., Calton, J. L., Hart, J. A., and Schachtman, T. R. (1999). Attenuation of the renewal effect by extinction in multiple contexts. *Learning and Motivation*, 30:1–14.

Chen, D. L., Moskowitz, T. J., and Shue, K. (2016). Decision making under the gambler's fallacy: Evidence from asylum judges, loan officers, and baseball umpires. *The Quarterly Journal of Economics*, 131:1181–1242.

Chiappori, P.-A., Levitt, S., and Groseclose, T. (2002). Testing mixed-strategy equilibria when players are heterogeneous: The case of penalty kicks in soccer. *American Economic Review*, 92:1138–1151.

Chinn, C. A. and Brewer, W. F. (1993). The role of anomalous data in knowledge acquisition: A theoretical framework and implications for science instruction. *Review of Educational Research*, 63:1–49.

Chomsky, N. (1965). *Aspects of the Theory of Syntax*. MIT Press.

Chomsky, N. (1986). *Knowledge of Language: Its Nature, Origin, and Use*. Greenwood Publishing Group.

Chomsky, N. (2002). *On Nature and Language*. Cambridge University Press.

Christy, A. G., Kim, J., Vess, M., Schlegel, R. J., and Hicks, J. A. (2017). The reciprocal relationship between perceptions of moral goodness and knowledge of others' true selves. *Social Psychological and Personality Science*, 8:910–917.

Chunharas, C., Rademaker, R. L., Brady, T., and Serences, J. (2019). Adaptive memory distortion in visual working memory. *PsyArXiv*. February 4, doi:10.31234/osf.io/e3m5a.

Churchland, M. M., Byron, M. Y., Cunningham, J. P., Sugrue, L. P., Cohen, M. R., Corrado, G. S., Newsome, W. T., Clark, A. M., Hosseini, P., Scott, B. B., et al. (2010). Stimulus onset quenches neural variability: A widespread cortical phenomenon. *Nature Neuroscience*, 13:369.

Churchland, P. M. (1979). *Scientific Realism and the Plasticity of Mind*. Cambridge University Press.

Clotfelter, C. T. and Cook, P. J. (1993). The "gambler's fallacy" in lottery play. *Management Science*, 39:1521–1525.

Coase, R. (1960). The problem of social cost. *The Journal of Law and Economics*, 3:1–44.

Cook, J. and Lewandowsky, S. (2016). Rational irrationality: Modeling climate change belief polarization using Bayesian networks. *Topics in Cognitive Science*, 8:160–179.

Cornsweet, T. (1970). *Visual Perception*. Academic Press.

Cover, T. M. and Thomas, J. A. (1991). *Elements of Information Theory*. John Wiley & Sons.

Cowan, J. W. (2016). *The Complete Lojban Language*. Logical Language Group.

Cox, R. T. (1946). Probability, frequency and reasonable expectation. *American Journal of Physics*, 14:1–13.

Craik, K. J. (1966). *The Nature of Psychology*. Cambridge University Press.

Cushman, F. (2019). Rationalization is rational. *Behavioral and Brain Sciences*, pages 1–69.

Dasgupta, I., Schulz, E., and Gershman, S. J. (2017). Where do hypotheses come from? *Cognitive Psychology*, 96:1–25.

Davis, D., Sundahl, I., and Lesbo, M. (2000). Illusory personal control as a determinant of bet size and type in casino craps games. *Journal of Applied Social Psychology*, 30:1224–1242.

Davis, E. and Marcus, G. (2015). Commonsense reasoning and commonsense knowledge in artificial intelligence. *Communications of the ACM*, 58:92–103.

De Finetti, B. (1931). Sul significato soggettivo della probabilita. *Fundamenta Mathematicae*, 17:298–329.

De Freitas, J., Sarkissian, H., Newman, G. E., Grossmann, I., De Brigard, F., Luco, A., and Knobe, J. (2018). Consistent belief in a good true self in misanthropes and three interdependent cultures. *Cognitive Science*, 42: 134–160.

De Freitas, J., Tobia, K. P., Newman, G. E., and Knobe, J. (2017). Normative judgments and individual essence. *Cognitive Science*, 41:382–402.

Deci, E. (1971). Effects of externally mediated rewards on intrinsic motivation. *Journal of Personality and Social Psychology*, 18:105–115.

Deci, E., Koestner, R., and Ryan, R. (1999). A meta-analytic review of experiments examining the effects of extrinsic rewards on intrinsic motivation. *Psychological Bulletin*, 125:627–668.

Deneve, S. (2008). Bayesian spiking neurons I: Inference. *Neural Computation*, 20:91–117.

Denison, S., Bonawitz, E., Gopnik, A., and Griffiths, T. L. (2013). Rational variability in children's causal inferences: The sampling hypothesis. *Cognition*, 126:285–300.

Dewar, K. M. and Xu, F. (2010). Induction, overhypothesis, and the origin of abstract knowledge: Evidence from 9-month-old infants. *Psychological Science*, 21:1871–1877.

Diaconis, P. (1978). Statistical problems in ESP research. *Science*, 201:131–136.

Dorfman, H. M., Bhui, R., Hughes, B. L., and Gershman, S. J. (2019). Causal inference about good and bad outcomes. *Psychological Science*, pages 516–525.

Dorling, J. (1979). Bayesian personalism, the methodology of scientific research programmes, and Duhem's problem. *Studies in History and Philosophy of Science Part A*, 10:177–187.

Dowman, M. (2007). Explaining color term typology with an evolutionary model. *Cognitive Science*, 31:99–132.

Doya, K., Ishii, S., Pouget, A., and Rao, R. P. (2007). *Bayesian Brain: Probabilistic Approaches to Neural Coding*. MIT Press.

Driver, P. M. and Humphries, D. A. (1988). *Protean Behaviour: The Biology of Unpredictability*. Clarendon Press.

Dubey, R., Agrawal, P., Pathak, D., Griffiths, T., and Efros, A. (2018). Investigating human priors for playing video games. In *International Conference on Machine Learning*, pages 1348–1356.

Duhem, P. M. (1954). *The Aim and Structure of Physical Theory*. Princeton University Press.

Duncker, K. (1929). Über induzierte bewegung. *Psychologische Forschung*, 12:180–259.

Dyson, G. (2012). *Turing's Cathedral: The Origins of the Digital Universe*. Pantheon.

Eagle, A. (2005). Randomness is unpredictability. *The British Journal for the Philosophy of Science*, 56:749–790.

Earman, J. (1992). *Bayes or Bust? A Critical Examination of Bayesian Confirmation Theory*. MIT Press.

Earman, J. (2000). *Hume's Abject Failure: The Argument Against Miracles*. Oxford University Press.

Eckhardt, R. (1987). Stan Ulam, John von Neumann, and the Monte Carlo method. *Los Alamos Science*, 15:131–136.

Eil, D. and Rao, J. M. (2011). The good news-bad news effect: Asymmetric processing of objective information about yourself. *American Economic Journal: Microeconomics*, 3:114–138.

Elster, J. (1987). *The Multiple Self*. Cambridge University Press.

Englich, B., Mussweiler, T., and Strack, F. (2006). Playing dice with criminal sentences: The influence of irrelevant anchors on experts' judicial decision making. *Personality and Social Psychology Bulletin*, 32:188–200.

Epley, N. and Gilovich, T. (2005). When effortful thinking influences judgmental anchoring: Differential effects of forewarning and incentives on self-generated and externally provided anchors. *Journal of Behavioral Decision Making*, 18:199–212.

Epley, N. and Gilovich, T. (2006). The anchoring-and-adjustment heuristic: Why the adjustments are insufficient. *Psychological Science*, 17:311–318.

Evans, J.S.B., Handley, S. J., Over, D. E., and Perham, N. (2002). Background beliefs in Bayesian inference. *Memory & Cognition*, 30:179–190.

Farnsworth, W. (2007). *The Legal Analyst: A Toolkit for Thinking About the Law*. The University of Chicago Press.

Fechner, G. T. (1860). *Elemente der Psychophysik*. Breitkopf und Härtel.

Fehr, E. and Gächter, S. (unpublished working paper). Do incentive contracts crowd out voluntary cooperation? IEW—Working Papers 034, Institute for Empirical Research in Economics, University of Zurich. See: RePEc:zur:iewwpx:034.

Fernandez-Duque, D. and Wifall, T. (2007). Actor/observer asymmetry in risky decision making. *Judgment and Decision Making*, 2:1.

Ferrell, W. R. and McGoey, P. J. (1980). A model of calibration for subjective probabilities. *Organizational Behavior and Human Performance*, 26:32–53.

Ferrucci, D., Brown, E., Chu-Carroll, J., Fan, J., Gondek, D., Kalyanpur, A. A., Lally, A., Murdock, J. W., Nyberg, E., Prager, J., et al. (2010). Building Watson: An overview of the DeepQA project. *AI Magazine*, 31:59–79.

Festinger, L. and Carlsmith, J. (1959). Cognitive consequences of forced compliance. *The Journal of Abnormal and Social Psychology*, 58:203–210.

Fitelson, B. and Waterman, A. (2005). Bayesian confirmation and auxiliary hypotheses revisited: A reply to Strevens. *The British Journal for the Philosophy of Science*, 56:293–302.

Flombaum, J. and Scholl, B. (2006). A temporal same-object advantage in the tunnel effect: Facilitated change detection for persisting objects. *Journal of Experimental Psychology: Human Perception and Performance*, 32:840–853.

Fodor, J. (1984). Observation reconsidered. *Philosophy of Science*, 51:23–43.

Frank, M. C. and Goodman, N. D. (2012). Predicting pragmatic reasoning in language games. *Science*, 336:998–998.

Frankfurt, H. (1971). Free will and the concept of a person. *Journal of Philosophy*, 68:5–20.

Freedman, J. and Fraser, S. (1966). Compliance without pressure. *Journal of Personality and Social Psychology*, 4:195–202.

Frey, B. S. and Jegen, R. (2001). Motivation crowding theory. *Journal of Economic Surveys*, 15:589–611.

Frey, B. S. and Oberholzer-Gee, F. (1997). The cost of price incentives: An empirical analysis of motivation crowding-out. *The American Economic Review*, 87:746–755.

Friston, K. (2005). A theory of cortical responses. *Philosophical Transactions of the Royal Society B: Biological Sciences*, 360:815–836.

Frosch, C. A. and Johnson-Laird, P. N. (2011). Is everyday causation deterministic or probabilistic? *Acta Psychologica*, 137:280–291.

Fudenberg, D. and Levine, D. K. (2006). A dual-self model of impulse control. *American Economic Review*, 96:1449–1476.

Fukushima, K. (1980). Neocognitron: A self-organizing neural network model for a mechanism of pattern recognition unaffected by shift in position. *Biological Cybernetics*, 36:193–202.

Furl, N. (2016). Facial-attractiveness choices are predicted by divisive normalization. *Psychological Science*, 27:1379–1387.

Futrell, R., Mahowald, K., and Gibson, E. (2015). Quantifying word order freedom in dependency corpora. In *Proceedings of the Third International Conference on Dependency Linguistics (Depling 2015)*, pages 91–100.

Gaier, A. and Ha, D. (2019). Weight agnostic neural networks. arXiv, Cornell University. arXiv:1906.04358.

Galinsky, A. and Mussweiler, T. (2001). First offers as anchors: The role of perspective-taking and negotiator focus. *Journal of Personality and Social Psychology*, 81:657–669.

Garrett, N. and Sharot, T. (2017). Optimistic update bias holds firm: Three tests of robustness following Shah et al. *Consciousness and Cognition*, 50:12–22.

Gell-Mann, M. and Ruhlen, M. (2011). The origin and evolution of word order. *Proceedings of the National Academy of Sciences*, 108:17290–17295.

George, D. (2009). *Preference Pollution: How Markets Create the Desires We Dislike*. University of Michigan Press.

Gershman, S. J. (2018). Deconstructing the human algorithms for exploration. *Cognition*, 173:34–42.

Gershman, S. J. (2019a). How to never be wrong. *Psychonomic Bulletin & Review*, 26:13–28.

Gershman, S. J. (2019b). Uncertainty and exploration. *Decision*, 6:277–286.

Gershman, S. J. and Beck, J. M. (2017). Complex probabilistic inference. In Moustafa, A., editor, *Computational Models of Brain and Behavior*, pages 453–466. Wiley-Blackwell.

Gershman, S. J., Horvitz, E. J., and Tenenbaum, J. B. (2015). Computational rationality: A converging paradigm for intelligence in brains, minds, and machines. *Science*, 349: 273–278.

Gershman, S. J., Jones, C. E., Norman, K. A., Monfils, M.-H., and Niv, Y. (2013). Gradual extinction prevents the return of fear: Implications for the discovery of state. *Frontiers in Behavioral Neuroscience*, 7:164.

Gershman, S. J., Monfils, M.-H., Norman, K. A., and Niv, Y. (2017a). The computational nature of memory modification. *eLife*, 6:e23763.

Gershman, S. J. and Niv, Y. (2013). Perceptual estimation obeys Occam's razor. *Frontiers in Psychology*, 4:623.

Gershman, S. J., Pouncy, H. T., and Gweon, H. (2017b). Learning the structure of social influence. *Cognitive Science*, 41:545–575.

Gershman, S. J., Radulescu, A., Norman, K. A., and Niv, Y. (2014). Statistical computations underlying the dynamics of memory updating. *PLoS Computational Biology*, 10:e1003939.

Gershman, S. J., Tenenbaum, J. B., and Jäkel, F. (2016). Discovering hierarchical motion structure. *Vision Research*, 126:232–241.

Gershman, S. J., Vul, E., and Tenenbaum, J. B. (2012). Multistability and perceptual inference. *Neural Computation*, 24:1–24.

Gerstenberg, T. and Tenenbaum, J. B. (2017). Intuitive theories. In Waldmann, M., editor, *Oxford Handbook of Causal Reasoning*. Oxford University Press.

Gervais, S. and Odean, T. (2001). Learning to be overconfident. *Review of Financial studies*, 14: 1–27.

Gibson, B. and Sanbonmatsu, D. M. (2004). Optimism, pessimism, and gambling: The downside of optimism. *Personality and Social Psychology Bulletin*, 30:149–160.

Gibson, E., Bergen, L., and Piantadosi, S. T. (2013a). Rational integration of noisy evidence and prior semantic expectations in sentence interpretation. *Proceedings of the National Academy of Sciences*, 110:8051–8056.

Gibson, E., Futrell, R., Jara-Ettinger, J., Mahowald, K., Bergen, L., Ratnasingam, S., Gibson, M., Piantadosi, S. T., and Conway, B. R. (2017). Color naming across languages reflects color use. *Proceedings of the National Academy of Sciences*, 114:10785–10790.

Gibson, E., Futrell, R., Piandadosi, S. T., Dautriche, I., Mahowald, K., Bergen, L., and Levy, R. (2019). How efficiency shapes human language. *Trends in Cognitive Sciences*, 23:389–407.

Gibson, E., Piantadosi, S. T., Brink, K., Bergen, L., Lim, E., and Saxe, R. (2013b). A noisy-channel account of crosslinguistic word-order variation. *Psychological Science*, 24:1079–1088.

Gigerenzer, G. and Gaissmaier, W. (2011). Heuristic decision making. *Annual Review of Psychology*, 62:451–482.

Gilbert, D. T. and Malone, P. S. (1995). The correspondence bias. *Psychological Bulletin*, 117: 21–38.

Gilovich, T. (1983). Biased evaluation and persistence in gambling. *Journal of Personality and Social Psychology*, 44:1110–1126.

Gilovich, T. (1991). *How We Know What Isn't So*. Simon and Schuster.

Gilovich, T. and Douglas, C. (1986). Biased evaluations of randomly determined gambling outcomes. *Journal of Experimental Social Psychology*, 22:228–241.

Gilovich, T., Vallone, R., and Tversky, A. (1985). The hot hand in basketball: On the misperception of random sequences. *Cognitive Psychology*, 17:295–314.

Gino, F., Norton, M. I., and Weber, R. A. (2016). Motivated Bayesians: Feeling moral while acting egoistically. *The Journal of Economic Perspectives*, 30:189–212.

Ginzel, L. E., Jones, E. E., and Swann Jr., W. B. (1987). How "naive" is the naive attributor? Discounting and augmentation in attitude attribution. *Social Cognition*, 5:108–130.

Gittins, J. C. (1979). Bandit processes and dynamic allocation indices. *Journal of the Royal Statistical Society: Series B (Methodological)*, 41:148–164.

Glimcher, P. W. (2005). Indeterminacy in brain and behavior. *Annual Review of Psychology*, 56:25–56.

Gneezy, A., Imas, A., Brown, A., Nelson, L. D., and Norton, M. I. (2012). Paying to be nice: Consistency and costly prosocial behavior. *Management Science*, 58:179–187.

Gneezy, U. and Potters, J. (1997). An experiment on risk taking and evaluation periods. *The Quarterly Journal of Economics*, 112:631–645.

Gneezy, U. and Rustichini, A. (2000a). A fine is a price. *The Journal of Legal Studies*, 29:1–17.

Gneezy, U. and Rustichini, A. (2000b). Pay enough or don't pay at all. *The Quarterly Journal of Economics*, 115:791–810.

Goertzel, T. (1994). Belief in conspiracy theories. *Political Psychology*, 15:731–742.

Goethals, G. R. and Reckman, R. F. (1973). The perception of consistency in attitudes. *Journal of Experimental Social Psychology*, 9:491–501.

Goldin-Meadow, S., So, W. C., Özyürek, A., and Mylander, C. (2008). The natural order of events: How speakers of different languages represent events nonverbally. *Proceedings of the National Academy of Sciences*, 105:9163–9168.

Goldstone, R. L. and Hendrickson, A. T. (2010). Categorical perception. *Wiley Interdisciplinary Reviews: Cognitive Science*, 1:69–78.

Golomb, J. D. (2015). Divided spatial attention and feature-mixing errors. *Attention, Perception, & Psychophysics*, 77:2562–2569.

Goodman, N. (1955). *Fact, Fiction, and Forecast*. Harvard University Press.

Goodman, N., Ullman, T., and Tenenbaum, J. (2011). Learning a theory of causality. *Psychological Review*, 118:110–119.

Goodman, N. D., Baker, C. L., and Tenenbaum, J. B. (2009). Cause and intent: Social reasoning in causal learning. In *Proceedings of the 31st Annual Conference of the Cognitive Science Society*, pages 2759–2764.

Goodman, N. D., Tenenbaum, J. B., Feldman, J., and Griffiths, T. L. (2008). A rational analysis of rule-based concept learning. *Cognitive Science*, 32:108–154.

Gopnik, A. (2012). Scientific thinking in young children: Theoretical advances, empirical research, and policy implications. *Science*, 337:1623–1627.

Gopnik, A. and Meltzoff, A. N. (1997). *Words, Thoughts, and Theories*. MIT Press.

Gopnik, A. and Slaughter, V. (1991). Young children's understanding of changes in their mental states. *Child Development*, 62:98–110.

Gopnik, A. and Wellman, H. M. (1992). Why the child's theory of mind really is a theory. *Mind & Language*, 7:145–171.

Greene, J. D., Sommerville, R. B., Nystrom, L. E., Darley, J. M., and Cohen, J. D. (2001). An fMRI investigation of emotional engagement in moral judgment. *Science*, 293:2105–2108.

Gregory, R. L. (1970). *The Intelligent Eye.* Weidenfeld and Nicolson.

Gregory, R. L. (1972). Cognitive contours. *Nature,* 238:51–52.

Grether, D. M. (1980). Bayes rule as a descriptive model: The representativeness heuristic. *The Quarterly Journal of Economics,* 95:537–557.

Grice, H. P. (1989). *Studies in the Way of Words.* Harvard University Press.

Griffin, D. and Tversky, A. (1992). The weighing of evidence and the determinants of confidence. *Cognitive Psychology,* 24:411–435.

Griffiths, T. L. and Tenenbaum, J. B. (2006). Optimal predictions in everyday cognition. *Psychological Science,* 17:767–773.

Griffiths, T. L. and Tenenbaum, J. B. (2007). From mere coincidences to meaningful discoveries. *Cognition,* 103:180–226.

Griffiths, T. L. and Tenenbaum, J. B. (2009). Theory-based causal induction. *Psychological Review,* 116:661–716.

Grinblatt, M. and Han, B. (2005). Prospect theory, mental accounting, and momentum. *Journal of Financial Economics,* 78:311–339.

Grünbaum, A. (1962). The falsifiability of theories: Total or partial? A contemporary evaluation of the Duhem-Quine thesis. *Synthese,* 14:17–34.

Gunther, L. M., Denniston, J. C., and Miller, R. R. (1998). Conducting exposure treatment in multiple contexts can prevent relapse. *Behaviour Research and Therapy,* 36:75–91.

Hacking, I. (1975). *The Emergence of Probability.* Cambridge University Press.

Hahn, M., Degen, J., Goodman, N., Jurafsky, D., and Futrell, R. (2018). An information-theoretic explanation of adjective ordering preferences. In *Proceedings of the 40th Annual Conference of the Cognitive Science Society.*

Hahn, U. and Oaksford, M. (2007). The rationality of informal argumentation. *Psychological Review,* 114:704–732.

Hahn, U., Oaksford, M., and Bayindir, H. (2005). How convinced should we be by negative evidence? In *Proceedings of the 27th Annual Conference of the Cognitive Science Society,* pages 887–892.

Hakes, J. K. and Sauer, R. D. (2006). An economic evaluation of the Moneyball hypothesis. *Journal of Economic Perspectives,* 20:173–186.

Hall, L., Johansson, P., Tärning, B., Sikström, S., and Deutgen, T. (2010). Magic at the marketplace: Choice blindness for the taste of jam and the smell of tea. *Cognition,* 117: 54–61.

Harding, S. (1976). *Can Theories Be Refuted?: Essays on the Duhem-Quine Thesis.* D. Reidel Publishing Company.

Harlow, H. (1949). The formation of learning sets. *Psychological Review,* 56:51–65.

Harris, A. J., Corner, A., and Hahn, U. (2013). James is polite and punctual (and useless): A Bayesian formalisation of faint praise. *Thinking & Reasoning,* 19:414–429.

Harris, A. J., Hsu, A. S., and Madsen, J. K. (2012). Because Hitler did it! Quantitative tests of Bayesian argumentation using ad hominem. *Thinking & Reasoning,* 18:311–343.

Harris, A. J. and Osman, M. (2012). The illusion of control: A Bayesian perspective. *Synthese,* 189:29–38.

Harris, P. (1996). Sufficient grounds for optimism? The relationship between perceived controllability and optimistic bias. *Journal of Social and Clinical Psychology,* 15:9–52.

Harvey, N. and Fischer, I. (1997). Taking advice: Accepting help, improving judgment, and sharing responsibility. *Organizational Behavior and Human Decision Processes*, 70:117–133.

Haslam, N., Bastian, B., and Bissett, M. (2004). Essentialist beliefs about personality and their implications. *Personality and Social Psychology Bulletin*, 30:1661–1673.

Haslam, S. A. and Reicher, S. D. (2008). Questioning the banality of evil. *Psychologist*, 21:16–19.

Hassabis, D., Kumaran, D., Summerfield, C., and Botvinick, M. (2017). Neuroscience-inspired artificial intelligence. *Neuron*, 95:245–258.

Hauser, M. D., Chomsky, N., and Fitch, W. T. (2002). The faculty of language: What is it, who has it, and how did it evolve? *Science*, 298:1569–1579.

Hawthorne-Madell, D. and Goodman, N. D. (2019). Reasoning about social sources to learn from actions and outcomes. *Decision*, 6:17–60.

Heider, F. and Simmel, M. (1944). An experimental study of apparent behavior. *The American Journal of Psychology*, 57:243–259.

Helmholtz, H. v. (1867). *Handbuch der physiologischen Optik*. Voss.

Hempel, C. G. (1966). *Philosophy of Natural Science*. Prentice-Hall.

Hendrickson, A. T., Navarro, D. J., and Perfors, A. (2016). Sensitivity to hypothesis size during information search. *Decision*, 3:62–80.

Henrich, J. (2017). *The Secret of Our Success: How Culture Is Driving Human Evolution, Domesticating Our Species, and Making Us Smarter*. Princeton University Press.

Hershfield, H. E., Goldstein, D. G., Sharpe, W. F., Fox, J., Yeykelis, L., Carstensen, L. L., and Bailenson, J. N. (2011). Increasing saving behavior through age-progressed renderings of the future self. *Journal of Marketing Research*, 48:S23–S37.

Hewstone, M. (1994). Revision and change of stereotypic beliefs: In search of the elusive subtyping model. *European Review of Social Psychology*, 5:69–109.

Hoffman, D. (2019). *The Case Against Reality: Why Evolution Hid the Truth from Our Eyes*. W. W. Norton & Company.

Horner, V. and Whiten, A. (2005). Causal knowledge and imitation/emulation switching in chimpanzees (*Pan troglodytes*) and children (*Homo sapiens*). *Animal Cognition*, 8:164–181.

Howson, C. (2011). *Objecting to God*. Cambridge University Press.

Howson, C. and Urbach, P. (2006). *Scientific Reasoning: The Bayesian Approach*. Open Court Publishing.

Hromádka, T., DeWeese, M. R., and Zador, A. M. (2008). Sparse representation of sounds in the unanesthetized auditory cortex. *PLoS Biology*, 6:e16.

Hsee, C. K. (1998). Less is better: When low-value options are valued more highly than high-value options. *Journal of Behavioral Decision Making*, 11:107–121.

Hsu, A. S. and Chater, N. (2010). The logical problem of language acquisition: A probabilistic perspective. *Cognitive Science*, 34:972–1016.

Hsu, A. S., Chater, N., and Vitányi, P. M. (2011). The probabilistic analysis of language acquisition: Theoretical, computational, and experimental analysis. *Cognition*, 120:380–390.

Huang, Y. and Rao, R. P. (2011). Predictive coding. *Wiley Interdisciplinary Reviews: Cognitive Science*, 2:580–593.

Huber, J., Payne, J. W., and Puto, C. (1982). Adding asymmetrically dominated alternatives: Violations of regularity and the similarity hypothesis. *Journal of Consumer Research*, 9: 90–98.

Hume, D. (1748). *An Enquiry Concerning Human Understanding.*

Huys, Q. J. and Dayan, P. (2009). A Bayesian formulation of behavioral control. *Cognition*, 113:314–328.

Icard, T. (2019). Why be random? *Mind.*

Jack, C. E. and Thurlow, W. R. (1973). Effects of degree of visual association and angle of displacement on the "ventriloquism" effect. *Perceptual and Motor Skills*, 37:967–979.

Janiszewski, C. and Lichtenstein, D. R. (1999). A range theory account of price perception. *Journal of Consumer Research*, 25:353–368.

Jaynes, E. T. (2003). *Probability Theory: The Logic of Science.* Cambridge University Press.

Jern, A., Chang, K.-M. K., and Kemp, C. (2014). Belief polarization is not always irrational. *Psychological Review*, 121:206–224.

Johansson, G. (1950). *Configurations in Event Perception.* Almqvist & Wiksell.

Johansson, P., Hall, L., Sikström, S., and Olsson, A. (2005). Failure to detect mismatches between intention and outcome in a simple decision task. *Science*, 310:116–119.

Johansson, P., Hall, L., Tärning, B., Sikström, S., and Chater, N. (2014). Choice blindness and preference change: You will like this paper better if you (believe you) chose to read it! *Journal of Behavioral Decision Making*, 27:281–289.

Johnson-Laird, P. N. (2006). *How We Reason.* Oxford University Press, USA.

Johnson-Laird, P. N. and Wason, P. C. (1970). A theoretical analysis of insight into a reasoning task. *Cognitive Psychology*, 1:134–148.

Johnston, L. and Hewstone, M. (1992). Cognitive models of stereotype change: 3. Subtyping and the perceived typicality of disconfirming group members. *Journal of Experimental Social Psychology*, 28:360–386.

Jones, E. E. and Harris, V. A. (1967). The attribution of attitudes. *Journal of Experimental Social Psychology*, 3:1–24.

Jones, M. and Love, B. C. (2011). Bayesian fundamentalism or enlightenment? On the explanatory status and theoretical contributions of Bayesian models of cognition. *Behavioral and Brain Sciences*, 34(4):169–188.

Juslin, P. (1993). An explanation of the hard-easy effect in studies of realism of confidence in one's general knowledge. *European Journal of Cognitive Psychology*, 5:55–71.

Kahneman, D. (2011). *Thinking, Fast and Slow.* Farrar, Straus and Giroux.

Kahneman, D., Knetsch, J. L., and Thaler, R. H. (1990). Experimental tests of the endowment effect and the Coase theorem. *Journal of Political Economy*, 98:1325–1348.

Kahneman, D. and Tversky, A. (1979). Prospect theory: An analysis of decision under risk. *Econometrica*, pages 263–291.

Kanizsa, G. (1955). Margini quasi-percettivi in campi con stimolazione omogenea. *Rivista di Psicologia*, 49:7–30.

Kao, S.-F. and Wasserman, E. (1993). Assessment of an information integration account of contingency judgment with examination of subjective cell importance and method of information presentation. *Journal of Experimental Psychology: Learning, Memory, and Cognition*, 19:1363–1386.

Kapadia, M. K., Westheimer, G., and Gilbert, C. D. (2000). Spatial distribution of contextual interactions in primary visual cortex and in visual perception. *Journal of Neurophysiology*, 84:2048–2062.

Karmiloff-Smith, A. and Inhelder, B. (1975). If you want to get ahead, get a theory. *Cognition*, 3:195–212.

Kaufman, L. and Rock, I. (1962). The Moon Illusion, I. *Science*, 136:953–961.

Kemeny, J. G. (1955). Fair bets and inductive probabilities. *The Journal of Symbolic Logic*, 20:263–273.

Kemp, C., Perfors, A., and Tenenbaum, J. B. (2007). Learning overhypotheses with hierarchical Bayesian models. *Developmental Science*, 10:307–321.

Kemp, C. and Tenenbaum, J. B. (2008). The discovery of structural form. *Proceedings of the National Academy of Sciences*, 105:10687–10692.

Kemp, C., Xu, Y., and Regier, T. (2018). Semantic typology and efficient communication. *Annual Review of Linguistics*, 4:109–128.

Kennedy, J. and Booth, C. (1963). Free flight of aphids in the laboratory. *Journal of Experimental Biology*, 40:67–85.

Kersten, D., Mamassian, P., and Yuille, A. (2004). Object perception as Bayesian inference. *Annual Review of Psychology*, 55:271–304.

Keupp, S., Bancken, C., Schillmöller, J., Rakoczy, H., and Behne, T. (2016). Rational over-imitation: Preschoolers consider material costs and copy causally irrelevant actions selectively. *Cognition*, 147:85–92.

Keupp, S., Behne, T., and Rakoczy, H. (2018). The rationality of (over) imitation. *Perspectives on Psychological Science*, 13:678–687.

Keupp, S., Behne, T., Zachow, J., Kasbohm, A., and Rakoczy, H. (2015). Over-imitation is not automatic: Context sensitivity in children's overimitation and action interpretation of causally irrelevant actions. *Journal of Experimental Child Psychology*, 130:163–175.

Kirby, K. N. (1994). Probabilities and utilities of fictional outcomes in Wason's four-card selection task. *Cognition*, 51:1–28.

Kirby, S., Cornish, H., and Smith, K. (2008). Cumulative cultural evolution in the laboratory: An experimental approach to the origins of structure in human language. *Proceedings of the National Academy of Sciences*, 105:10681–10686.

Kirby, S., Dowman, M., and Griffiths, T. L. (2007). Innateness and culture in the evolution of language. *Proceedings of the National Academy of Sciences*, 104:5241–5245.

Kirby, S., Tamariz, M., Cornish, H., and Smith, K. (2015). Compression and communication in the cultural evolution of linguistic structure. *Cognition*, 141:87–102.

Klayman, J. and Ha, Y.-W. (1987). Confirmation, disconfirmation, and information in hypothesis testing. *Psychological Review*, 94:211–228.

Kleinmuntz, B. (1990). Why we still use our heads instead of formulas: Toward an integrative approach. *Psychological Bulletin*, 107:296–310.

Kliegl, R., Grabner, E., Rolfs, M., and Engbert, R. (2004). Length, frequency, and predictability effects of words on eye movements in reading. *European Journal of Cognitive Psychology*, 16:262–284.

Knapen, T., Kanai, R., Brascamp, J., van Boxtel, J., and van Ee, R. (2007). Distance in feature space determines exclusivity in visual rivalry. *Vision Research*, 47:3269–3275.

Knill, D. C. and Kersten, D. (1991). Apparent surface curvature affects lightness perception. *Nature*, 351(6323):228.

Knill, D. C. and Richards, W. (1996). *Perception as Bayesian Inference*. Cambridge University Press.

Knudsen, V. O. (1923). The sensibility of the ear to small differences of intensity and frequency. *Physical Review*, 21:84.

Koehler, J. J. (1993). The influence of prior beliefs on scientific judgments of evidence quality. *Organizational Behavior and Human Decision Processes*, 56:28–55.

Koehler, J. J. and Conley, C. A. (2003). The "hot hand" myth in professional basketball. *Journal of Sport and Exercise Psychology*, 25:253–259.

Kohn, A. (1993). *Punished by Rewards*. Plenum Press.

Kolers, P. and Pomerantz, J. (1971). Figural change in apparent motion. *Journal of Experimental Psychology*, 87:99–108.

Konkle, T. and Oliva, A. (2011). Canonical visual size for real-world objects. *Journal of Experimental Psychology: Human Perception and Performance*, 37:23–37.

Kool, W., Gershman, S. J., and Cushman, F. A. (2017). Cost-benefit arbitration between multiple reinforcement-learning systems. *Psychological Science*, 28:1321–1333.

Kool, W., McGuire, J., Rosen, Z., and Botvinick, M. (2010). Decision making and the avoidance of cognitive demand. *Journal of Experimental Psychology: General*, 139:665–682.

Körding, K. P., Beierholm, U., Ma, W. J., Quartz, S., Tenenbaum, J. B., and Shams, L. (2007). Causal inference in multisensory perception. *PLoS One*, 2:e943.

Korn, C., Sharot, T., Walter, H., Heekeren, H., and Dolan, R. (2014). Depression is related to an absence of optimistically biased belief updating about future life events. *Psychological Medicine*, 44:579–592.

Korn, C. W., Prehn, K., Park, S. Q., Walter, H., and Heekeren, H. R. (2012). Positively biased processing of self-relevant social feedback. *Journal of Neuroscience*, 32:16832–16844.

Köszegi, B. (2006). Ego utility, overconfidence, and task choice. *Journal of the European Economic Association*, 4:673–707.

Kuhn, T. S. (1962). *The Structure of Scientific Revolutions*. University of Chicago Press.

Kunda, Z. (1990). The case for motivated reasoning. *Psychological Bulletin*, 108:480–498.

Kuzmanovic, B. and Rigoux, L. (2017). Valence-dependent belief updating: Computational validation. *Frontiers in Psychology*, 8:1087.

Ladouceur, R. and Sévigny, S. (2005). Structural characteristics of video lotteries: Effects of a stopping device on illusion of control and gambling persistence. *Journal of Gambling Studies*, 21:117–131.

Lakatos, I. (1976). Falsification and the methodology of scientific research programmes. In *Can Theories Be Refuted?*, pages 205–259. Springer.

Lake, B. M., Salakhutdinov, R., and Tenenbaum, J. B. (2015). Human-level concept learning through probabilistic program induction. *Science*, 350:1332–1338.

Lake, B. M., Ullman, T. D., Tenenbaum, J. B., and Gershman, S. J. (2017). Building machines that learn and think like people. *Behavioral and Brain Sciences*, 40.

Landau, B., Smith, L. B., and Jones, S. S. (1988). The importance of shape in early lexical learning. *Cognitive Development*, 3:299–321.

Langer, E. J. (1975). The illusion of control. *Journal of personality and social psychology*, 32: 311–328.

Lau, T., Pouncy, H., Gershman, S., and Cikara, M. (2018). Discovering social groups via latent structure learning. *Journal of Experimental Psychology: General*, 147:1881–1891.

Laudan, L. (1990). Demystifying underdetermination. *Minnesota Studies in the Philosophy of Science*, 14(1990):267–297.

Laughlin, S. B. (2001). Energy as a constraint on the coding and processing of sensory information. *Current Opinion in Neurobiology*, 11:475–480.

Lazear, E. P. (2000). Performance pay and productivity. *American Economic Review*, 90:1346–1361.

Lee, S.-H., Blake, R., and Heeger, D. J. (2005). Traveling waves of activity in primary visual cortex during binocular rivalry. *Nature Neuroscience*, 8:22–23.

Lee, T. S. and Yuille, A. L. (2006). Efficient coding of visual scenes by grouping and segmentation. In Doya, K., Ishii, S., Pouget, A., and Rao, R. P., editors, *Bayesian Brain: Probabilistic Approaches to Neural Coding*, pages 141–185. MIT Press.

Lefebvre, G., Lebreton, M., Meyniel, F., Bourgeois-Gironde, S., and Palminteri, S. (2017). Behavioural and neural characterization of optimistic reinforcement learning. *Nature Human Behaviour*, 1:0067.

Lehman, R. S. (1955). On confirmation and rational betting. *The Journal of Symbolic Logic*, 20:251–262.

Leiserowitz, A. A., Maibach, E. W., Roser-Renouf, C., Smith, N., and Dawson, E. (2013). Climategate, public opinion, and the loss of trust. *American Behavioral Scientist*, 57:818–837.

Lepper, M., Greene, D., and Nisbett, R. (1973). Undermining children's intrinsic interest with extrinsic reward. *Journal of Personality and Social Psychology*, 28:129–137.

Leslie, A. M. and Keeble, S. (1987). Do six-month-old infants perceive causality? *Cognition*, 25:265–288.

Lewandowsky, S., Griffiths, T. L., and Kalish, M. L. (2009). The wisdom of individuals: Exploring people's knowledge about everyday events using iterated learning. *Cognitive Science*, 33:969–998.

Lewandowsky, S., Oberauer, K., and Gignac, G. E. (2013). NASA faked the moon landing—therefore, (climate) science is a hoax: An anatomy of the motivated rejection of science. *Psychological Science*, 24:622–633.

Lichtenstein, S. and Slovic, P. (2006). *The Construction of Preference*. Cambridge University Press.

Lieder, F. and Griffiths, T. L. (2019). Resource-rational analysis: Understanding human cognition as the optimal use of limited computational resources. *Behavioral and Brain Sciences*, pages 1–85.

Lieder, F., Griffiths, T. L., Huys, Q. J., and Goodman, N. D. (2018a). The anchoring bias reflects rational use of cognitive resources. *Psychonomic Bulletin & Review*, 25:322–349.

Lieder, F., Griffiths, T. L., Huys, Q. J., and Goodman, N. D. (2018b). Empirical evidence for resource-rational anchoring and adjustment. *Psychonomic Bulletin & Review*, 25:775–784.

Lindley, D. V. et al. (1956). On a measure of the information provided by an experiment. *The Annals of Mathematical Statistics*, 27:986–1005.

Lippmann, W. (1922). *Public Opinion*. Harcourt, Brace.

List, J. A. (2002). Preference reversals of a different kind: The "more is less" phenomenon. *American Economic Review*, 92:1636–1643.

Loewenstein, G. (1987). Anticipation and the valuation of delayed consumption. *The Economic Journal*, 97:666–684.

Loomis, J. and Nakayama, K. (1973). A velocity analogue of brightness contrast. *Perception*, 2:425–428.

Lord, C. G., Ross, L., and Lepper, M. R. (1979). Biased assimilation and attitude polarization: The effects of prior theories on subsequently considered evidence. *Journal of Personality and Social Psychology*, 37:2098–2109.

Louie, K., Khaw, M. W., and Glimcher, P. W. (2013). Normalization is a general neural mechanism for context-dependent decision making. *Proceedings of the National Academy of Sciences*, 110:6139–6144.

Lu, H., Yuille, A. L., Liljeholm, M., Cheng, P. W., and Holyoak, K. J. (2008). Bayesian generic priors for causal learning. *Psychological review*, 115:955–984.

Lupyan, G. and Dale, R. (2016). Why are there different languages? The role of adaptation in linguistic diversity. *Trends in Cognitive Sciences*, 20:649–660.

Luttmer, E. F. (2005). Neighbors as negatives: Relative earnings and well-being. *The Quarterly Journal of Economics*, 120:963–1002.

Lyons, D. E., Santos, L. R., and Keil, F. C. (2006). Reflections of other minds: How primate social cognition can inform the function of mirror neurons. *Current Opinion in Neurobiology*, 16:230–234.

Ma, W. J., Beck, J. M., Latham, P. E., and Pouget, A. (2006). Bayesian inference with probabilistic population codes. *Nature Neuroscience*, 9:1432–1438.

MacKay, D. J. (2003). *Information Theory, Inference and Learning Algorithms*. Cambridge University Press.

Mainen, Z. F. and Sejnowski, T. J. (1995). Reliability of spike timing in neocortical neurons. *Science*, 268:1503–1506.

Mandel, D. and Lehman, D. (1998). Integration of contingency information in judgments of cause, covariation, and probability. *Journal of Experimental Psychology: General*, 127:269–285.

Mandelbrot, B. (1953). An informational theory of the statistical structure of language. In Jackson, W., editor, *Communication Theory*, pages 486–502. Betterworth.

Maratsos, M. (2000). More overregularizations after all: New data and discussion on Marcus, Pinker, Ullman, Hollander, Rosen & Xu. *Journal of Child Language*, 27:183–212.

Marcus, G. (2009). *Kluge: The Haphazard Evolution of the Human Mind*. Houghton Mifflin Harcourt.

Marcus, G. F. and Davis, E. (2013). How robust are probabilistic models of higher-level cognition? *Psychological Science*, 24:2351–2360.

Marcus, G. F., Pinker, S., Ullman, M., Hollander, M., Rosen, T. J., Xu, F., and Clahsen, H. (1992). Overregularization in language acquisition. *Monographs of the Society for Research in Child Development*, pages i–178.

Markman, E. M. and Hutchinson, J. E. (1984). Children's sensitivity to constraints on word meaning: Taxonomic versus thematic relations. *Cognitive Psychology*, 16:1–27.

Mayo, D. G. (1997). Duhem's problem, the Bayesian way, and error statistics, or "What's belief got to do with it?" *Philosophy of Science*, 64:222–244.

Mayrhofer, R. and Waldmann, M. R. (2015). Sufficiency and necessity assumptions in causal structure induction. *Cognitive Science*, 40:2137–2150.

McAllester, D. A. (1999). Some PAC-Bayesian theorems. *Machine Learning*, 37:355–363.

McBrayer, J. P. (2010). Skeptical theism. *Philosophy Compass*, 5:611–623.

McClelland, J. L. and Patterson, K. (2002). 'Words or rules' cannot exploit the regularity in exceptions. *Trends in Cognitive Sciences*, 6:464–464.

McGuigan, N., Makinson, J., and Whiten, A. (2011). From over-imitation to super-copying: Adults imitate causally irrelevant aspects of tool use with higher fidelity than young children. *British Journal of Psychology*, 102:1–18.

McGuigan, N. and Whiten, A. (2009). Emulation and "overemulation" in the social learning of causally opaque versus causally transparent tool use by 23- and 30-month-olds. *Journal of Experimental Child Psychology*, 104:367–381.

McGuigan, N., Whiten, A., Flynn, E., and Horner, V. (2007). Imitation of causally opaque versus causally transparent tool use by 3- and 5-year-old children. *Cognitive Development*, 22:353–364.

McKenzie, C. R., Ferreira, V. S., Mikkelsen, L. A., McDermott, K. J., and Skrable, R. P. (2001). Do conditional hypotheses target rare events? *Organizational Behavior and Human Decision Processes*, 85:291–309.

Mckenzie, C. R. and Mikkelsen, L. A. (2000). The psychological side of Hempel's paradox of confirmation. *Psychonomic Bulletin & Review*, 7:360–366.

McKenzie, C. R. and Mikkelsen, L. A. (2007). A Bayesian view of covariation assessment. *Cognitive Psychology*, 54:33–61.

McNamara, J. (1982). *Names for Things: A Study of Child Language*. MIT Press.

Mehra, R. and Prescott, E. C. (1985). The equity premium: A puzzle. *Journal of Monetary Economics*, 15:145–161.

Mellers, B. A., Schwartz, A., Ho, K., and Ritov, I. (1997). Decision affect theory: Emotional reactions to the outcomes of risky options. *Psychological Science*, 8:423–429.

Mercier, H. and Sperber, D. (2017). *The Enigma of Reason*. Harvard University Press.

Merritt, A. C., Effron, D. A., Fein, S., Savitsky, K. K., Tuller, D. M., and Monin, B. (2012). The strategic pursuit of moral credentials. *Journal of Experimental Social Psychology*, 48:774–777.

Michotte, A. (1946). *The Perception of Causality*. Basic Books.

Milgram, S. (1974). *Obedience to Authority: An Experimental View*. Harper Collins.

Milkman, K. L., Rogers, T., and Bazerman, M. H. (2010). I'll have the ice cream soon and the vegetables later: A study of online grocery purchases and order lead time. *Marketing Letters*, 21:17–35.

Miller, G. F. (1997). Protean primates: The evolution of adaptive unpredictability in competition and courtship. *Machiavellian Intelligence II: Extensions and Evaluations*, 2:312.

Miller, J. B. and Sanjurjo, A. (2018). Surprised by the hot hand fallacy? A truth in the law of small numbers. *Econometrica*, 86, 2019–2047.

Molouki, S. and Bartels, D. M. (2017). Personal change and the continuity of the self. *Cognitive Psychology*, 93:1–17.

Monin, B. and Miller, D. (2001). Moral credentials and the expression of prejudice. *Journal of Personality and Social Psychology*, 81:33–43.

Monroe, D. (2014). Neuromorphic computing gets ready for the (really) big time. *Communications of the ACM*, 57:13–15.

Moore, D. and Healy, P. (2008). The trouble with overconfidence. *Psychological Review*, 115: 502–517.

Moore, D. and Small, D. (2007). Error and bias in comparative judgment. *Journal of Personality and Social Psychology*, 92:972–989.

Moore, M. T. and Fresco, D. M. (2012). Depressive realism: A meta-analytic review. *Clinical Psychology Review*, 32:496–509.

Mora, T. and Bialek, W. (2011). Are biological systems poised at criticality? *Journal of Statistical Physics*, 144:268–302.

Moravcik, M., Schmid, M., Burch, N., Lisỳ, V., Morrill, D., Bard, N., Davis, T., Waugh, K., Johanson, M., and Bowling, M. (2017). Deepstack: Expert-level artificial intelligence in heads-up no-limit poker. *Science*, 356:508–513.

Morgenstern, L., Davis, E., and Ortiz, C. L. (2016). Planning, executing, and evaluating the Winograd schema challenge. *AI Magazine*, 37:50–54.

Morris, M. W. and Larrick, R. P. (1995). When one cause casts doubt on another: A normative analysis of discounting in causal attribution. *Psychological Review*, 102:331–355.

Morse, S. and Gergen, K. (1970). Social comparison, self-consistency, and the concept of self. *Journal of Personality and Social Psychology*, 16:148–156.

Mozer, M. C., Pashler, H., and Homaei, H. (2008). Optimal predictions in everyday cognition: The wisdom of individuals or crowds? *Cognitive Science*, 32:1133–1147.

Muentener, P. and Schulz, L. (2014). Toddlers infer unobserved causes for spontaneous events. *Frontiers in Psychology*, 5.

Mussweiler, T. (2003). Comparison processes in social judgment: Mechanisms and consequences. *Psychological Review*, 110:472–489.

Navarro, D. J. and Perfors, A. F. (2011). Hypothesis generation, sparse categories, and the positive test strategy. *Psychological Review*, 118:120–134.

Nawrot, M. and Sekuler, R. (1990). Assimilation and contrast in motion perception: Explorations in cooperativity. *Vision Research*, 30:1439–1451.

Nelson, M. E. and Bower, J. M. (1990). Brain maps and parallel computers. *Trends in Neurosciences*, 13:403–408.

Newman, G. E., Bloom, P., and Knobe, J. (2014). Value judgments and the true self. *Personality and Social Psychology Bulletin*, 40:203–216.

Newman, G. E., De Freitas, J., and Knobe, J. (2015). Beliefs about the true self explain asymmetries based on moral judgment. *Cognitive Science*, 39:96–125.

Newman, M. E. (2005). Power laws, Pareto distributions and Zipf's law. *Contemporary Physics*, 46:323–351.

Nickerson, R. S. (1998). Confirmation bias: A ubiquitous phenomenon in many guises. *Review of General Psychology*, 2:175–220.

Nieder, A. and Miller, E. K. (2003). Coding of cognitive magnitude: Compressed scaling of numerical information in the primate prefrontal cortex. *Neuron*, 37:149–157.

Nisbett, R. and Wilson, T. (1977). Telling more than we can know: Verbal reports on mental processes. *Psychological Review*, 84:231–259.

Nisbett, R. E. and Ross, L. (1980). *Human Inference: Strategies and Shortcomings of Social Judgment*. Prentice-Hall.

Nosofsky, R., Palmeri, T., and McKinley, S. (1994). Rule-plus-exception model of classification learning. *Psychological Review*, 101:53–79.

Nowak, M. A., Komarova, N. L., and Niyogi, P. (2002). Computational and evolutionary aspects of language. *Nature*, 417:611–617.

Oaksford, M. and Chater, N. (1994). A rational analysis of the selection task as optimal data selection. *Psychological Review*, 101:608–631.

Oaksford, M. and Chater, N. (2007). *Bayesian Rationality: The Probabilistic Approach to Human Reasoning*. Oxford University Press.

Oaksford, M. and Hahn, U. (2004). A Bayesian approach to the argument from ignorance. *Canadian Journal of Experimental Psychology*, 58:75–85.

O'Brien, V. (1959). Contrast by contour-enhancement. *American Journal of Psychology*, 72:299.

Odean, T. (1998). Are investors reluctant to realize their losses? *The Journal of Finance*, 53:1775–1798.

Okrent, A. (2009). *In the Land of Invented Languages: Esperanto Rock Stars, Klingon Poets, Loglan Lovers, and the Mad Dreamers Who Tried to Build a Perfect Language.* Spiegel & Grau.

Olshausen, B. A. and Field, D. J. (1996). Emergence of simple-cell receptive field properties by learning a sparse code for natural images. *Nature*, 381:607–609.

Orbán, G., Berkes, P., Fiser, J., and Lengyel, M. (2016). Neural variability and sampling-based probabilistic representations in the visual cortex. *Neuron*, 92:530–543.

O'Shea, R. P., Sims, A. J., and Govan, D. G. (1997). The effect of spatial frequency and field size on the spread of exclusive visibility in binocular rivalry. *Vision Research*, 37:175–183.

Palacios-Huerta, I. (2003). Professionals play minimax. *The Review of Economic Studies*, 70:395–415.

Parducci, A. (1965). Category judgment: A range-frequency model. *Psychological Review*, 72:407–418.

Parducci, A. and Perrett, L. (1971). Category rating scales. *Journal of Experimental Psychology*, 89:427–452.

Parducci, A. and Wedell, D. (1986). The category effect with rating scales. *Journal of Experimental Psychology: Human Perception and Performance*, 12:496–516.

Parfit, D. (1984). *Reasons and Persons*. Oxford University Press.

Paul, L. A. (2014). *Transformative Experience*. Oxford University Press.

Pecevski, D., Buesing, L., and Maass, W. (2011). Probabilistic inference in general graphical models through sampling in stochastic networks of spiking neurons. *PLoS Computational Biology*, 7:e1002294.

Perfors, A. and Navarro, D. (2009). Confirmation bias is rational when hypotheses are sparse. *Proceedings of the 31st Annual Conference of the Cognitive Science Society.*

Perner, J., Leekam, S. R., and Wimmer, H. (1987). Three-year-olds' difficulty with false belief: The case for a conceptual deficit. *British Journal of Developmental Psychology*, 5:125–137.

Peterson, C. (1980). Recognition of noncontingency. *Journal of Personality and Social Psychology*, 38:727–734.

Piantadosi, S. T. and Jacobs, R. A. (2016). Four problems solved by the probabilistic language of thought. *Current Directions in Psychological Science*, 25:54–59.

Piantadosi, S. T., Palmeri, H., and Aslin, R. (2018). Limits on composition of conceptual operations in 9-month-olds. *Infancy*, 23:310–324.

Piantadosi, S. T., Tily, H., and Gibson, E. (2012). The communicative function of ambiguity in language. *Cognition*, 122:280–291.

Pierson, E. and Goodman, N. (2014). Uncertainty and denial: A resource-rational model of the value of information. *PloS One*, 9:e113342.

Pinker, S. and Ullman, M. T. (2002). The past and future of the past tense. *Trends in Cognitive Sciences*, 6:456–463.

Pollack, I. (1952). The information of elementary auditory displays. *The Journal of the Acoustical Society of America*, 24:745–749.

Pollack, I. (1953). The information of elementary auditory displays. II. *The Journal of the Acoustical Society of America*, 25:765–769.

Polo, M. (1918). *The Travels of Marco Polo*. J. M. Dent & Sons.

Poo, C. and Isaacson, J. S. (2009). Odor representations in olfactory cortex: "Sparse" coding, global inhibition, and oscillations. *Neuron*, 62:850–861.

Popper, K. (1959). *The Logic of Scientific Discovery*. Harper & Row.

Powell, D., Merrick, M. A., Lu, H., and Holyoak, K. J. (2016). Causal competition based on generic priors. *Cognitive Psychology*, 86:62–86.

Prasada, S. and Pinker, S. (1993). Generalisation of regular and irregular morphological patterns. *Language and Cognitive Processes*, 8:1–56.

Prelec, D., Wernerfelt, B., and Zettelmeyer, F. (1997). The role of inference in context effects: Inferring what you want from what is available. *Journal of Consumer Research*, 24: 118–125.

Prendergast, C. (1999). The provision of incentives in firms. *Journal of Economic Literature*, 37:7–63.

Qian, T. and Jaeger, T. F. (2012). Cue effectiveness in communicatively efficient discourse production. *Cognitive Science*, 36:1312–1336.

Quattrone, G. and Tversky, A. (1984). Causal versus diagnostic contingencies: On self-deception and on the voter's illusion. *Journal of Personality and Social Psychology*, 46: 237–248.

Queller, S. and Smith, E. R. (2002). Subtyping versus bookkeeping in stereotype learning and change: Connectionist simulations and empirical findings. *Journal of Personality and Social Psychology*, 82:300–313.

Quine, W. V. (1951). Two dogmas of empiricism. *The Philosophical Review*, pages 20–43.

Rabin, M. and Vayanos, D. (2010). The gambler's and hot-hand fallacies: Theory and applications. *The Review of Economic Studies*, 77:730–778.

Rapoport, A. and Budescu, D. (1992). Generation of random series in two-person strictly competitive games. *Journal of Experimental Psychology: General*, 121:352–363.

Ravazzolo, F. and Røisland, Ø. (2011). Why do people place lower weight on advice far from their own initial opinion? *Economics Letters*, 112:63–66.

Regier, T., Carstensen, A., and Kemp, C. (2016). Languages support efficient communication about the environment: Words for snow revisited. *PloS One*, 11:e0151138.

Regier, T. and Kay, P. (2009). Language, thought, and color: Whorf was half right. *Trends in Cognitive Sciences*, 13:439–446.

Regier, T., Kemp, C., and Kay, P. (2015). Word meanings across languages support efficient communication. In *The Handbook of Language Emergence*, pages 237–264. Wiley.

Reiter, R. (1980). A logic for default reasoning. *Artificial Intelligence*, 13:81–132.

Rescorla, R. A. (2004). Spontaneous recovery. *Learning & Memory*, 11:501–509.

Restle, F. (1970). Moon illusion explained on the basis of relative size. *Science*, 167:1092–1096.

Roberson, D., Davidoff, J., Davies, I. R., and Shapiro, L. R. (2005). Color categories: Evidence for the cultural relativity hypothesis. *Cognitive Psychology*, 50:378–411.

Robert, C. (2007). *The Bayesian Choice: From Decision-theoretic Foundations to Computational Implementation*. Springer Science & Business Media.

Robert, C. and Casella, G. (2013). *Monte Carlo Statistical Methods*. Springer Science & Business Media.

Rock, I. (1983). *The Logic of Perception*. MIT Press.

Rock, I. and Kaufman, L. (1962). The moon illusion, II: The moon's apparent size is a function of the presence or absence of terrain. *Science*, (3521):1023–1031.

Roeder, K. D. (1962). The behaviour of free flying moths in the presence of artificial ultrasonic pulses. *Animal Behaviour*, 10:300–304.

Rosch, E. (1978). Principles of categorization. In *Cognition and Categorization*, pages 27–48. Lawrence Erbaum Associates.

Rutchick, A., Slepian, M., Reyes, M., Pleskus, L., and Hershfield, H. (2018). Future self-continuity is associated with improved health and increases exercise behavior. *Journal of Experimental Psychology: Applied*, 24:72–80.

Rutledge, R. B., Skandali, N., Dayan, P., and Dolan, R. J. (2014). A computational and neural model of momentary subjective well-being. *Proceedings of the National Academy of Sciences*, 111:12252–12257.

Saxe, R., Tenenbaum, J., and Carey, S. (2005). Secret agents: Inferences about hidden causes by 10- and 12-month-old infants. *Psychological Science*, 16:995–1001.

Schachner, A. and Carey, S. (2013). Reasoning about 'irrational' actions: When intentional movements cannot be explained, the movements themselves are seen as the goal. *Cognition*, 129:309–327.

Schank, R. C. (1972). Conceptual dependency: A theory of natural language understanding. *Cognitive Psychology*, 3:552–631.

Schelling, T. C. (1978). Egonomics, or the art of self-management. *The American Economic Review*, 68:290–294.

Schmidt, L. A., Goodman, N. D., Barner, D., and Tenenbaum, J. B. (2009). How tall is tall? Compositionality, statistics, and gradable adjectives. In *Proceedings of the 31st Annual Conference of the Cognitive Science Society*, pages 2759–2764.

Scholl, B. J. and Pylyshyn, Z. W. (1999). Tracking multiple items through occlusion: Clues to visual objecthood. *Cognitive Psychology*, 38:259–290.

Schroeder, M. (1991). *Fractals, Chaos, Power Laws: Minutes from an Infinte Paradise*. Freeman.

Schultz, W., Dayan, P., and Montague, P. R. (1997). A neural substrate of prediction and reward. *Science*, 275:1593–1599.

Schulz, E., Bhui, R., Love, B. C., Brier, B., Todd, M. T., and Gershman, S. J. (2019). Structured, uncertainty-driven exploration in real-world consumer choice. *Proceedings of the National Academy of Sciences*, 16:13903–13908.

Schulz, L. E., Goodman, N. D., Tenenbaum, J. B., and Jenkins, A. C. (2008). Going beyond the evidence: Abstract laws and preschoolers' responses to anomalous data. *Cognition*, 109:211–223.

Schulz, L. E. and Sommerville, J. (2006). God does not play dice: Causal determinism and preschoolers' causal inferences. *Child Development*, 77:427–442.

Schustack, M. and Sternberg, R. (1981). Evaluation of evidence in causal inference. *Journal of Experimental Psychology: General*, 110:101–120.

Scontras, G., Degen, J., and Goodman, N. D. (2017). Subjectivity predicts adjective ordering preferences. *Open Mind*, 1:53–66.

Seligman, M. E. (1975). *Helplessness: On Depression, Development, and Death.* W. H. Freeman/ Times Books/Henry Holt & Co.

Shaffer, R. and Jadwiszczok, A. (2016). Psychic defective: Sylvia Browne's history of failure. *Skeptical Inquirer*, 34:38–42.

Shah, P., Harris, A. J., Bird, G., Catmur, C., and Hahn, U. (2016). A pessimistic view of optimistic belief updating. *Cognitive Psychology*, 90:71–127.

Shannon, C. E. and Weaver, W. (1949). *The Mathematical Theory of Communication.* University of Illinois Press.

Sharot, T. (2011). *The Optimism Bias.* Vintage.

Sharot, T. and Garrett, N. (2016). Forming beliefs: Why valence matters. *Trends in Cognitive Sciences*, 20:25–33.

Sharot, T., Korn, C. W., and Dolan, R. J. (2011). How unrealistic optimism is maintained in the face of reality. *Nature Neuroscience*, 14:1475–1479.

Shefrin, H. and Statman, M. (1985). The disposition to sell winners too early and ride losers too long: Theory and evidence. *The Journal of Finance*, 40:777–790.

Shenoy, P. and Yu, A. (2013). Rational preference shifts in multi-attribute choice: What is fair? In *Proceedings of the Annual Meeting of the Cognitive Science Society*, 35:1300–1305.

Sher, S. and McKenzie, C. (2014). Options as information: Rational reversals of evaluation and preference. *Journal of Experimental Psychology: General*, 143:1127–1143.

Shipp, T. D., Shipp, D. Z., Bromley, B., Sheahan, R., Cohen, A., Lieberman, E., and Benacerraf, B. (2004). What factors are associated with parents' desire to know the sex of their unborn child? *Birth*, 31:272–279.

Silver, D., Huang, A., Maddison, C. J., Guez, A., Sifre, L., Van Den Driessche, G., Schrittwieser, J., Antonoglou, I., Panneershelvam, V., Lanctot, M., et al. (2016). Mastering the game of Go with deep neural networks and tree search. *Nature*, 529:484.

Silver, D., Hubert, T., Schrittwieser, J., Antonoglou, I., Lai, M., Guez, A., Lanctot, M., Sifre, L., Kumaran, D., Graepel, T., et al. (2018). A general reinforcement learning algorithm that masters chess, shogi, and Go through self-play. *Science*, 362:1140–1144.

Simons, D. J. and Levin, D. T. (1998). Failure to detect changes to people during a real-world interaction. *Psychonomic Bulletin & Review*, 5:644–649.

Simonson, I. (1989). Choice based on reasons: The case of attraction and compromise effects. *Journal of Consumer Research*, 16:158–174.

Slutsky, D. A. and Recanzone, G. H. (2001). Temporal and spatial dependency of the ventriloquism effect. *Neuroreport*, 12:7–10.

Smith, L. B., Jones, S. S., Landau, B., Gershkoff-Stowe, L., and Samuelson, L. (2002). Object name learning provides on-the-job training for attention. *Psychological Science*, 13:13–19.

Spelke, E., Breinlinger, K., Macomber, J., and Jacobson, K. (1992). Origins of knowledge. *Psychological Review*, 99:605–632.

Spelke, E. S. (1990). Principles of object perception. *Cognitive Science*, 14:29–56.

Srinivasan, M. V., Laughlin, S. B., and Dubs, A. (1982). Predictive coding: A fresh view of inhibition in the retina. *Proceedings of the Royal Society of London. Series B. Biological Sciences*, 216:427–459.

Stankevicius, A., Huys, Q. J., Kalra, A., and Seriès, P. (2014). Optimism as a prior belief about the probability of future reward. *PLoS Computational Biology*, 10:e1003605.

Stanley, M. L., Henne, P., and De Brigard, F. (2019). Remembering moral and immoral actions in constructing the self. *Memory & Cognition*, 47:441–454.

Stevens, S. (1957). On the psychophysical law. *Psychological Review*, 64:153–181.

Stewart, N., Chater, N., and Brown, G. D. (2006). Decision by sampling. *Cognitive Psychology*, 53:1–26.

Stewart, N., Reimers, S., and Harris, A. J. (2014). On the origin of utility, weighting, and discounting functions: How they get their shapes and how to change their shapes. *Management Science*, 61:687–705.

Strandberg, T., Sivén, D., Hall, L., Johansson, P., and Pärnamets, P. (2018). False beliefs and confabulation can lead to lasting changes in political attitudes. *Journal of Experimental Psychology: General*, 147:1382–1399.

Stratton, G. M. (1897). Vision without inversion of the retinal image. *Psychological Review*, 4:341–360.

Strauss, L. (1953). *Natural Right and History*. University of Chicago Press.

Strevens, M. (2001). The Bayesian treatment of auxiliary hypotheses. *The British Journal for the Philosophy of Science*, 52:515–537.

Strohminger, N., Knobe, J., and Newman, G. (2017). The true self: A psychological concept distinct from the self. *Perspectives on Psychological Science*. 12:551–560.

Strotz, R. H. (1955). Myopia and inconsistency in dynamic utility maximization. *The Review of Economic Studies*, 23:165–180.

Suetens, S., Galbo-Jørgensen, C. B., and Tyran, J.-R. (2016). Predicting lotto numbers: A natural experiment on the gambler's fallacy and the hot-hand fallacy. *Journal of the European Economic Association*, 14:584–607.

Sunstein, C. R. and Vermeule, A. (2009). Conspiracy theories: Causes and cures. *Journal of Political Philosophy*, 17:202–227.

Sutton, R. S. and Barto, A. G. (2018). *Reinforcement Learning: An Introduction*. MIT Press.

Sweeny, K., Melnyk, D., Miller, W., and Shepperd, J. A. (2010). Information avoidance: Who, what, when, and why. *Review of General Psychology*, 14:340–353.

Swinburne, R. G. (1970). *The Concept of Miracle*. Springer.

Swinburne, R. G. (2004). *The Existence of God*. Oxford University Press.

Tauber, S., Navarro, D., Perfors, A., and Steyvers, M. (2017). Bayesian models of cognition revisited: Setting optimality aside and letting data drive psychological theory. *Psychological Review*, 124:410–441.

Tetlock, P., Kristel, O., Elson, S., Green, M., and Lerner, J. (2000). The psychology of the unthinkable. *Journal of Personality and Social Psychology*, 78:853–870.

Thaker, P., Tenenbaum, J. B., and Gershman, S. J. (2017). Online learning of symbolic concepts. *Journal of Mathematical Psychology*, 77:10–20.

Thaler, R. H. and Shefrin, H. M. (1981). An economic theory of self-control. *Journal of Political Economy*, 89:392–406.

Thaler, R. H., Tversky, A., Kahneman, D., and Schwartz, A. (1997). The effect of myopia and loss aversion on risk taking: An experimental test. *The Quarterly Journal of Economics*, 112:647–661.

Titmuss, R. (1970). *The Gift Relationship*. Allen and Unwin.

Todorov, E. (2004). Optimality principles in sensorimotor control. *Nature Neuroscience*, 7:907–915.

Tolhurst, D. J., Movshon, J. A., and Dean, A. F. (1983). The statistical reliability of signals in single neurons in cat and monkey visual cortex. *Vision Research*, 23:775–785.

Tomov, M. S., Truong, V. Q., Hundia, R. A., and Gershman, S. J. (2020). Dissociable neural correlates of uncertainty underlie different exploration strategies. *Nature Communications*, 11:1–12.

Tsividis, P. A., Pouncy, T., Xu, J. L., Tenenbaum, J. B., and Gershman, S. J. (2017). Human learning in Atari. In *2017 AAAI Spring Symposium Series*.

Tucker, J., Vuchinich, R., and Sobell, M. (1981). Alcohol consumption as a self-handicapping strategy. *Journal of Abnormal Psychology*, 90:220–230.

Tversky, A. (1972). Elimination by aspects. *Psychological Review*, 79:281–299.

Tversky, A. and Kahneman, D. (1974). Judgment under uncertainty: Heuristics and biases. *Science*, 185:1124–1131.

Tversky, A. and Shafir, E. (1992). The disjunction effect in choice under uncertainty. *Psychological Science*, 3:305–310.

Valenza, E., Leo, I., Gava, L., and Simion, F. (2006). Perceptual completion in newborn human infants. *Child Development*, 77:1810–1821.

Van Gelder, J.-L., Hershfield, H. E., and Nordgren, L. F. (2013). Vividness of the future self predicts delinquency. *Psychological Science*, 24:974–980.

Van Gelder, J.-L., Luciano, E. C., Weulen Kranenbarg, M., and Hershfield, H. E. (2015). Friends with my future self: Longitudinal vividness intervention reduces delinquency. *Criminology*, 53:158–179.

Van Rooy, D., Van Overwalle, F., Vanhoomissen, T., Labiouse, C., and French, R. (2003). A recurrent connectionist model of group biases. *Psychological Review*, 110:536–563.

Varga, C. A. (2001). Coping with HIV/AIDS in Durban's commercial sex industry. *AIDS Care*, 13:351–365.

Vohs, K. D., Baumeister, R. F., and Chin, J. (2007). Feeling duped: Emotional, motivational, and cognitive aspects of being exploited by others. *Review of General Psychology*, 11: 127–141.

Von Neumann, J. and Morgenstern, O. (1944). *Theory of Games and Economic Behavior*. Princeton University Press.

Vosniadou, S. and Brewer, W. F. (1992). Mental models of the earth: A study of conceptual change in childhood. *Cognitive Psychology*, 24:535–585.

Voss, R. and Clarke, J. (1975). '1/f noise' in music and speech. *Nature*, 258:317–318.

Vul, E., Goodman, N., Griffiths, T. L., and Tenenbaum, J. B. (2014). One and done? Optimal decisions from very few samples. *Cognitive Science*, 38:599–637.

Wagenaar, W. (1972). Generation of random sequences by human subjects: A critical survey of literature. *Psychological Bulletin*, 77:65–72.

Wagenmakers, E.-J., Wetzels, R., Borsboom, D., and van der Maas, H. (2011). Why psychologists must change the way they analyze their data: The case of psi: Comment on Bem (2011). *Journal of Personality and Social Psychology*, 100:426–432.

Walasek, L. and Stewart, N. (2015). How to make loss aversion disappear and reverse: Tests of the decision by sampling origin of loss aversion. *Journal of Experimental Psychology: General*, 144:7–11.

Walker, D., Smith, K. A., and Vul, E. (2015). The 'fundamental attribution error' is rational in an uncertain world. In *Proceedings of the Cognitive Science Society*.

Walker, M. and Wooders, J. (2001). Minimax play at wimbledon. *American Economic Review*, 91:1521–1538.

Wallace, M. T., Roberson, G., Hairston, W. D., Stein, B. E., Vaughan, J. W., and Schirillo, J. A. (2004). Unifying multisensory signals across time and space. *Experimental Brain Research*, 158:252–258.

Walton, D. (1996). *Arguments from Ignorance*. Pennsylvania State University Press.

Warren, D. H., Welch, R. B., and McCarthy, T. J. (1981). The role of visual-auditory "compellingness" in the ventriloquism effect: Implications for transitivity among the spatial senses. *Perception & Psychophysics*, 30:557–564.

Wason, P. and Green, D. (1984). Reasoning and mental representation. *The Quarterly Journal of Experimental Psychology Section A*, 36:597–610.

Wason, P. C. (1968). Reasoning about a rule. *Quarterly Journal of Experimental Psychology*, 20:273–281.

Wason, P. C. (1969). Regression in reasoning? *British Journal of Psychology*, 60:471–480.

Weber, R. and Crocker, J. (1983). Cognitive processes in the revision of stereotypic beliefs. *Journal of Personality and Social Psychology*, 45:961–977.

Wedell, D. H., Parducci, A., and Geiselman, R. E. (1987). A formal analysis of ratings of physical attractiveness: Successive contrast and simultaneous assimilation. *Journal of Experimental Social Psychology*, 23:230–249.

Wei, X.-X. and Stocker, A. A. (2015). A Bayesian observer model constrained by efficient coding can explain "anti-Bayesian" percepts. *Nature Neuroscience*, 18:1509.

Weinstein, N. D. (1980). Unrealistic optimism about future life events. *Journal of Personality and Social Psychology*, 39:806–820.

Welch, I. (1992). Sequential sales, learning, and cascades. *The Journal of Finance*, 47:695–732.

Werch, C. E. and Owen, D. M. (2002). Iatrogenic effects of alcohol and drug prevention programs. *Journal of Studies on Alcohol*, 63:581–590.

Werner, G. and Mountcastle, V. B. (1965). Neural activity in mechanoreceptive cutaneous afferents: Stimulus-response relations, Weber functions, and information transmission. *Journal of Neurophysiology*, 28:359–397.

Wernerfelt, B. (1995). A rational reconstruction of the compromise effect: Using market data to infer utilities. *Journal of Consumer Research*, 21:627–633.

Wertheimer, M. (1912). Experimentelle studien uber das sehen von bewegung. *Zeitschrift fur Psychologie*, 61.

Westbrook, A., Kester, D., and Braver, T. S. (2013). What is the subjective cost of cognitive effort? Load, trait, and aging effects revealed by economic preference. *PloS one*, 8: e68210.

Whiten, A., Allan, G., Devlin, S., Kseib, N., Raw, N., and McGuigan, N. (2016). Social learning in the real-world: 'Over-imitation' occurs in both children and adults unaware of participation in an experiment and independently of social interaction. *PLoS One*, 11:e0159920.

Whittington, J. C. and Bogacz, R. (2019). Theories of error back-propagation in the brain. *Trends in Cognitive Sciences*, 23:235–250.

Wilson, H. R., Blake, R., and Lee, S.-H. (2001). Dynamics of travelling waves in visual perception. *Nature*, 412:907–910.

Wilson, T. D. (2004). *Strangers to Ourselves*. Harvard University Press.

Wimmer, H. and Perner, J. (1983). Beliefs about beliefs: Representation and constraining function of wrong beliefs in young children's understanding of deception. *Cognition*, 13:103–128.

Winograd, T. (1972). Understanding natural language. *Cognitive Psychology*, 3:1–191.

Witkin, A. P. and Tenenbaum, J. M. (1983). On the role of structure in vision. In *Human and Machine Vision*, pages 481–543. Elsevier.

Wu, C.-C. and Chen, C.-C. (2018). The effect of size statistics of the background texture on perceived target size. *Scientific Reports*, 8:10963.

Wu, Y., Muentener, P., and Schulz, L. E. (2015). The invisible hand: Toddlers connect probabilistic events with agentive causes. *Cognitive Science*, 40:1854–1876.

Yamins, D. L. and DiCarlo, J. J. (2016). Using goal-driven deep learning models to understand sensory cortex. *Nature Neuroscience*, 19:356–365.

Yaniv, I. (2004). Receiving other people's advice: Influence and benefit. *Organizational Behavior and Human Decision Processes*, 93:1–13.

Yaniv, I. and Kleinberger, E. (2000). Advice taking in decision making: Egocentric discounting and reputation formation. *Organizational Behavior and Human Decision Processes*, 83: 260–281.

Yeung, S. and Griffiths, T. L. (2015). Identifying expectations about the strength of causal relationships. *Cognitive Psychology*, 76:1–29.

Yuille, A. and Kersten, D. (2006). Vision as Bayesian inference: Analysis by synthesis? *Trends in Cognitive Sciences*, 10:301–308.

Zador, A. M. (2019). A critique of pure learning and what artificial neural networks can learn from animal brains. *Nature Communications*, 10:1–7.

Zaslavsky, N., Kemp, C., Regier, T., and Tishby, N. (2018). Efficient compression in color naming and its evolution. *Proceedings of the National Academy of Sciences*, 115:7937–7942.

Zipf, G. K. (1949). *Human Behavior and the Principle of Least Effort*. Addison-Wesley.

INDEX